Sustainable Carbon Neutrality

Chunyan Chen · Maosheng Duan · Zhiyuan Guo ·
Binbin Zhang · Qian Wang · Hongli Wei

Sustainable Carbon Neutrality

中国工信出版集团 電子工業出版社 PUBLISHING HOUSE OF ELECTRONICS INDUSTRY http://www.phei.com.cn

Springer

Chunyan Chen
Guangzhou Ceprei Certification Center Services Limited
Guangzhou, China

Zhiyuan Guo
Guangzhou Ceprei Certification Center Services Limited
Guangzhou, China

Qian Wang
Guangzhou Ceprei Certification Center Services Limited
Guangzhou, China

Maosheng Duan
Institute of Energy, Environment and Economy
Tsinghua University
Beijing, China

Binbin Zhang
China Electronic Product Reliability and Environmental Testing Research Institute
Guangzhou, China

Hongli Wei
Guangzhou Ceprei Certification Center Services Limited
Guangzhou, China

ISBN 978-981-95-1690-2 ISBN 978-981-95-1691-9 (eBook)
https://doi.org/10.1007/978-981-95-1691-9

Jointly published with Publishing House of Electronics Industry
The print edition is not for sale in China (Mainland). Customers from China (Mainland) please order the print book from: Publishing House of Electronics Industry.

This Springer imprint is published by the registered company Springer Nature Singapore Pte Ltd.
The registered company address is: 152 Beach Road, #21-01/04 Gateway East, Singapore 189721, Singapore

Preface

As the challenges of global climate change become increasingly prominent, achieving carbon neutrality has become a goal and responsibility shared by the globe. The international community has reached consensus on the issues of climate change and adopted the Paris Agreement and the carbon neutrality goal. Although the signatory countries have made efforts in climate governance, a gap still remains in meeting the goals of the Paris Agreement. China has clearly put forward the "Dual Carbon" goals (carbon peaking by 2030 and carbon neutrality by 2060), which signifies not only a long-term plan for domestic economic and social development, but also an important contribution to global climate actions. Carbon neutrality is not only an environmental issue, but also a comprehensive challenge involving multiple fields such as economy, society, technology, and policy. The authors of this book, having long been committed to the research and practice on climate change and energy consumption, feel it is necessary to share the latest progress in achieving carbon neutrality with the public and readers to promote wider public understanding and participation.

This book aims to provide a comprehensive perspective to help readers understand the scientific basis, policy frameworks, technical pathways and practical cases of carbon neutrality. There are seven chapters in this book, each of which revolves around a specific topic. Starting from the global carbon neutrality goal and some carbon neutrality practices in China, Chap. 1 mainly introduces the basic knowledge and policy status of carbon neutrality at home and abroad. Chapter 2 analyzes the key issues of the pathways to achieve carbon neutrality in depth and reveals the core ideas of carbon neutrality. The subsequent chapters introduce in detail the key pathways to achieve sustainable carbon neutrality at the regional, organizational and product levels. Following the basic framework of carbon neutrality commitment—carbon emission reduction—carbon offsetting, it focuses on the key technologies for low-carbon transformation in key areas, carbon sinks, and carbon capture, utilization and storage, etc., striving to reveal the effective ways and strategies to achieve carbon neutrality.

The journey towards "carbon neutrality" presents both challenges and opportunities, which will bring about a major transformation of the economy and society and a major change in a wide range of fields. Achieving carbon neutrality means ensuring, through various means, that human activities no longer produce a net increase in the concentration of carbon dioxide in the atmosphere. The exploration of sustainable carbon neutrality pathways is a multidimensional and interdisciplinary field involving technological progress, economic development, policy formulation, and international cooperation. Achieving these goals requires the joint efforts of the whole society, including the participation of governments, enterprises, research institutions, and the public.

This book is suitable for all readers who are interested in carbon neutrality, including policy makers, corporate decision makers, academic researchers, students and the general public. Whether you are an expert or a beginner in this field, this book can provide you with useful information and inspiration. It is the hope of the authors that readers can gain knowledge and inspiration from this book, and that it can be a good starting point for you to explore the world of carbon neutrality.

Guangzhou, China Chunyan Chen

Introduction

Starting from the global carbon neutrality goal and some of China's Dual Carbon Policy (3060 goals) and practices, this book analyzes the key factors in the pathways toward carbon neutrality, and introduces in detail the key pathways to achieve sustainable carbon neutrality at the regional, organizational and product levels. Following the basic framework of carbon neutrality commitment-carbon emission reduction-carbon offsetting, it focuses on the key technologies for low-carbon transformation in key areas, Carbon Capture, Utilization and Storage (CCUS), and carbon sinks, etc. This book delivers actionable guidance for carbon neutrality practitioners from governments, enterprises and research institutions in clarifying the concept of carbon neutrality, quantifying emissions, formulating emission reduction plans and achieving carbon neutrality. This book is suitable for all readers who are interested in the carbon neutrality topic, including policy makers, corporate decision makers, academic researchers, students and the general public.

Contents

Chapter 1
Carbon Neutrality Overview

1 Global Temperature Control and Carbon Neutrality Goals

1.1 The Paris Agreement and Carbon Neutrality

The Paris Agreement, a legally binding international treaty with historic meaning, was adopted by 196 Parties at the 21st Session of the Conference of the Parties to the United Nations Framework Convention on Climate Change (UNFCCC) in Paris, France, on December 12, 2015. The Paris Agreement sets the goal to hold the increase in the global average temperature to well below 2 °C above pre-industrial levels and pursue efforts to limit the temperature increase to 1.5 °C above pre-industrial levels by the end of the twenty-first century. It also innovates the way in which the parties set their targets, that is, each party submits its Nationally Determined Contribution (NDC) goals and action plans, and is pushed to continuously enhance the strength of their commitments and actions through a global stocktake every five years. Meanwhile, the Paris Agreement requires achieving a balance between anthropogenic emissions by sources and removals by sinks of greenhouse gases in the second half of the twenty-first century.

The Paris Agreement provides a framework for global response to climate change after 2020. It is an important milestone in global climate governance, changing the mechanism of mandatory allocating emission reduction responsibilities from a higher level to a lower one and taking global climate governance into a new model of communicating the needs from a lower level up. This agreement gathers the consensus of mankind to protect the Earth as our home, takes into account the economic and social differences between countries. It provides the basic guidelines for international actions to address climate change after 2020.

This section mainly elaborates on the reasons why the cause of carbon neutrality is necessary. It will first sort out the process of international climate negotiations and

C. Chen et al., *Sustainable Carbon Neutrality*,
https://doi.org/10.1007/978-981-95-1691-9_1

the milestone events, and expound on how the international community has gradually reached a consensus on climate change and adopted the Paris Agreement as well as the carbon neutrality target. Secondly, it will illustrate that climate problems exist and are related to human activities. Although all parties to the convention have made efforts in climate governance, there is still a gap between these efforts and the goals of the Paris Agreement. Therefore, it is necessary for both state and non-state actors to intensify their efforts in emission reduction actions.

a. Review of the International Negotiations on Climate Change

Throughout the history of international negotiations on climate change, there are roughly five stages in the development of international climate cooperation and legislation and rules.

Phase I (1992–1997), the one for the United Nations Framework Convention on Climate Change (UNFCCC). The UNFCCC adopted in 1992 established the global goals for addressing climate change and clarified principles such as "Common But Differentiated Responsibilities (CBDR-RC, per UNFCCC Article 3.1)". It required developed countries to take the lead in action taking and to provide financial and technological assistance to developing countries. The UNFCCC officially initiated the international cooperation process for addressing climate change, and the international community was optimistic about cooperating to deal with climate change.

Phase II (1997–2005), the one for the Kyoto Protocol(2005–2020, superseded by Paris Agreement) which unilaterally stipulated the emission reduction obligations for developed countries. The Kyoto Protocol further refined the specific goals and mechanisms for implementation, including clarifying the overall goal for developed countries to reduce emissions by 5.2% below 1990 levels during the first commitment period (from 2008 to 2012), and establishing three flexible mechanisms, namely the International Emissions Trading (IET), the Joint Implementation (JI) and the Clean Development Mechanism (CDM). However, the United States refused to sign this agreement. During this stage, although China, as a developing country, did not undertake mandatory emission reduction obligations, it needed to assume the general obligation to control greenhouse gases and also assist in the implementation of the CDM established in the Kyoto Protocol. The fact that the Kyoto Protocol only set quantified emission reduction obligations for developed country parties is of historical significance, but it also became an important cause for criticism from the international community. This led to stronger demands from developed countries for developing countries to strengthen their emission reduction actions.

Phase III (2005–2010), the one for the "Bali Roadmap" in which the "dual-track" negotiations were launched. The "dual-track" refers to that, on the one hand, the developed countries that signed the Kyoto Protocol were required to fulfill the provisions of the Kyoto Protocol and commit to substantial quantified emission reduction targets after 2012; on the other hand, developing countries and developed countries that did not sign the Kyoto Protocol (mainly referring to the United States) were required to take further measures to address climate change under the UNFCCC. The "dual-track" emphasized the "shared responsibilities" under the principle of "common but differentiated responsibilities", made it clear that developing countries also had to take actions within their capabilities, and preliminarily established the "bottom-up" model of "commitment + review". According to the decisions in the "Bali Roadmap", the Copenhagen Protocol was supposed to be adopted in 2009 to replace the Kyoto Protocol which would expire in 2012. However, the Copenhagen Protocol was ultimately not accepted, and after that, the legal binding force of the Cancun Agreements adopted in 2010 was not satisfactory either.

Phase IV (2011–2019), the one for the "Durban Platform" and the Paris Agreement. From the launch of the Durban Platform for Enhanced Action in 2011, through the adoption of the Doha Amendment to the Kyoto Protocol in 2012, and then to the landmark Paris Agreement reached in 2015, which is applicable to all parties to the UNFCCC, the cooperation models and mechanisms for the international response to climate change were further developed. These include clarifying the goal of reaching the peak of emissions globally as early as possible, adding the goal of pursuing efforts to limit the temperature increase to 1.5 °C through the first round of periodic reviews, establishing the Green Climate Fund (GCF), proposing the Internationally Transferred Mitigation Outcomes (ITMO) and the Sustainable Development Mechanism (SDM), etc.

The Paris Agreement is the second binding climate agreement under the UNFCCC following the Kyoto Protocol. It has made a balanced arrangement for various elements such as mitigation, adaptation, finance, technology, capacity building, transparency and the global stocktake. It requires all parties to make Nationally Determined Contributions (NDCs), which has changed the emission reduction model of the Kyoto Protocol where quantified emission reduction targets were set for developed countries through negotiations while exempting developing countries. The new mitigation mechanism established by the Paris Agreement is mainly reflected in the following aspects: establishing the global long-term goal of "limiting global temperature increase to below 2 °C, while pursuing efforts to limit the increase to 1.5 °C", and adhering to the principle of "common but differentiated responsibilities" to achieve a win–win situation for multiple parties based on the Intended Nationally Determined Contributions (INDCs).

Meanwhile, in order to respond to the 2 and 1.5 °C targets and make them more achievable, the Intergovernmental Panel on Climate Change (IPCC) conducted a large number of studies on the emission reduction pathways to achieve the 2 °C target in its Fifth Assessment Report released in 2014. Subsequently, in response to the requirements of the UNFCCC, it further released the Special Report on Global Warming of 1.5 °C in 2018. The Special Report on Global Warming of 1.5 °C proposed that the 1.5 °C target is technically feasible, can significantly reduce climate risks and negative impacts compared to a 2 °C temperature rise, and is generally conducive to achieving the United Nations Sustainable Development Goals (SDGs) by 2030. However, achieving the 1.5 °C target requires a more urgent emission reduction process, with global carbon emissions reaching their peak as soon as possible and then starting to decline rapidly, and other greenhouse gases achieving deep emission reductions. Moreover, its cost is also much higher than that of achieving the 2 °C target.

Phase V (2020–Present), the one for the full implementation of the Paris Agreement. The Paris Agreement sets the goal of achieving a balance between anthropogenic emissions by sources and removals by sinks of greenhouse gases in the second half of the twenty-first century, that is, the global carbon neutrality goal. The Paris Agreement has made arrangements for the global action plans to address climate change after 2020 and represents an important turning point in the global governance models and concepts. Meanwhile, 2020 was a crucial year when the parties agreed to update their NDCs and report their low-greenhouse gas emission development strategies for 2050. The Paris Agreement stipulates that governments will take stock of the implementation of the Agreement periodically to assess the collective progress towards achieving the purpose of the Agreement and its long-term goals (referred to as the global stocktake). The first stocktake was in 2023, and then it would be conducted every five years. The outcome of the stocktake will show the progress of global emission reductions and the gaps between the INDC goals of the countries and the emission scenarios for achieving the global long-term goals, providing information for all parties to further motivate them to update and strengthen their INDC goals, actions and funding efforts, and to allow increased international cooperation.

In the future, the international community will continuously intensify its actions in the cooperation in addressing climate change, set more ambicious goals for NDCs, and jointly make greater contributions to the global process of addressing climate change.

b. It is Urgent for Both State and Non-state Actors to Work Together Towards the Goals of the Paris Agreement

Although all parties jointly adopted the global temperature control goals in the Paris Agreement, however, according to the reports of many authoritative international organizations and academic research, the current climate problems are severe and there is a major gap between the actual global temperature increase data and the goals of the Paris Agreement. Therefore, all relevant parties, including state and non-state actors, need to continue to make greater efforts.

Firstly, climate problems are severe and humans need to take responsibility for them. In March 2023, the IPCC finalized the Synthesis Report of the Sixth Assessment Report (AR6), namely "AR6 Synthesis Report: Climate Change 2023", which pointed out that human activities, principally through greenhouse gas emissions, have unequivocally caused global warming, and that the intensified global warming will bring multiple hazards. The "State of the Global Climate 2022" report released by the World Meteorological Organization (WMO) also indicated that the global mean temperature in 2022 was 1.15 ± 0.13 °C above the pre-industrial average from 1850 to 1900, and it is very likely the years 2015 to 2022 are the eight warmest years on record.

Secondly, all parties have made efforts regarding climate issues, but these efforts are still insufficient. The *Emissions Gap Report 2022: The Closing Window – Climate crisis calls for rapid transformation of societies* released by the United Nations Environment Programme (UNEP) pointed out that the impacts of climate change are becoming increasingly severe on a global scale. However, the international community is falling far short of the goals of the Paris Agreement at present. Only an urgent system-wide transformation can avoid climate disaster. The AR6 Synthesis Report of IPCC also believes that, with high confidence, the best estimate is that the global temperature rise will reach 1.5 °C in a short term, and there is a gap between the projected emissions of the implemented policies and those of the Nationally Determined Contributions (NDCs). Financially, it has not reached the level that enables all sectors and regions to achieve climate goals. The academic community has also drawn similar conclusions. A study in 2023 pointed out that even if the commitments made by countries at the 26th Conference of the Parties (COP26) to the UNFCCC in 2021 are fulfilled, it can only limit the global temperature increase mean between 2.4 °C and 2.8 °C, making it difficult to meet the goals of the Paris Agreement. Fortunately, the international community has also intensified its efforts to address climate change in recent years, which has alleviated the deterioration of the temperature issue to a certain extent and also demonstrated the effectiveness of emission reduction actions. For example, based on the latest *2022 NDC Synthesis Report* published by the UNFCCC, 74% of the parties that submitted or updated their NDCs have strengthened their commitments to reducing or limiting greenhouse gas emissions by 2025 or 2030, indicating that countries have responded to the relevant calls of the UNFCCC and demonstrated greater climate ambitions. Moreover, considering the implementation of the latest NDCs, global emissions may peak before 2030, and the possibility of peaking is higher than that in the previous report (the 2016 report). Certainly, the report also pointed out that although the INDCs have increased, according to the latest NDCs and IPCC scenarios, the absolute difference in the emission levels by 2030 is still significant. This shows that, since the COP26 in 2021, all parties have boosted their ambitions to address climate change and attached greater importance to climate issues. However, the international community still needs more comprehensive emission reduction actions.

Based on the previous two points, it can be concluded that if the goals of the Paris Agreement are to be achieved, it is urgently necessary for both state and non-state actors to adopt effective measures all together. More countries need to formulate long-term strategies that are aligned with the Paris Agreement. Moreover, the updated NDCs of each country need to be consistent with the net-zero emission goals. The countries should follow the goals and principles of the Paris Agreement, work more on international cooperation actions, encourage all parties to put in their own actions more effort, significantly strengthen and update their respective NDC goals and actions, inspire countries to increase their ambitions to address climate change, and promote the comprehensive, balanced and effective implementation of the Paris Agreement.

In addition to the actions of state actors, the actions of non-state actors are also of great importance. Non-state actors refer to various non-governmental organizations, think tanks, scientific research institutions and religious groups from civil society, as well as companies and investors from the private sector, and substate actors such as cities and regions. The actions of non-state actors usually refer to "a series of governance actions carried out outside the strict governmental and inter-governmental (or multilateral) environment". The non-state actors' involvement in international climate negotiations and climate governance has gone through three major stages, which are the embryonic stage (1992–2008) characterized by their participation in negotiations, the stage (2009–2013) when their influence inside and outside negotiations rapidly increased, and the stage after 2014 when actions outside negotiations were incorporated into the formal process of the UNFCCC. Besides actively participating in climate negotiations, non-state actors also take actions in a wide range of fields such as renewable energy, forests and agriculture, energy efficiency, transportation, climate adaptation and resilience, climate financing, and infrastructure, and excert an influence on climate mitigation and adaptation in multiple ways. Currently, non-state actors are not only promoters, supervisors and reviewers of national climate policies, but also implementers of climate governance, which has increased the possibility of achieving the carbon neutrality goal to a certain extent. Of course, non-state actors can achieve carbon neutrality at a regional, organizational, project and even product level below the national level.

1.2 Diversified Carbon Neutrality Goals

As understood from the previous section, the carbon neutrality goals need to be jointly driven by state actors and non-state actors at multiple levels. Therefore, this section will introduce in detail the different levels of carbon neutrality goals respectively. Carbon neutrality goals are diverse in terms of the implementing actors and implementation levels, and can be specifically divided into four levels: carbon neutrality at the regional level, carbon neutrality at the organizational level, carbon neutrality at the project level and carbon neutrality at the product level. There are certain differences among the four in aspects such as the scope of accounting, accounting standards and accounting methods.

a. Carbon Neutrality at the Regional Level

Carbon neutrality at the regional level is based on the geographical scope, usually a country or a region, and it calculates the total greenhouse gas emissions absorbed by all producers within it. As per its definition, carbon neutrality at the regional level can be further defined as carbon neutrality at the national-level regional level and carbon neutrality at the sub-national regional level.

The carbon neutrality goals at the national-level regional level are the "main force" in pushing the achievement of the overall carbon neutrality goal globally. Each government puts forward their carbon neutrality commitments and formulate and publish relevant policies and laws. In the recent years, a lot of countries and regions have successively announced their "carbon neutrality" goals, laying a foundation for achieving the goals of the Paris Agreement. According to the carbon neutrality tracking dataset published by the University of Oxford, by the end of

October 2023, 6 countries worldwide had already achieved carbon neutrality, and 87 countries and regions (including the European Union) had announced carbon neutrality commitments or included them in national policies or laws. Additionally, although 58 countries and regions had not yet made formal commitments, they had already proposed or were discussing carbon neutrality goals. Table 1 divides the countries in the world that have made substantial commitments and progress regarding carbon neutrality goals into four categories according to the degree of progress from high to low, namely, countries that have achieved carbon neutrality, countries that have legislated for carbon neutrality, countries that have included carbon neutrality goals in policies, and countries that have formally announced or committed to carbon neutrality, and lists the countries under the each category. For countries that have not yet achieved carbon neutrality, Table 1 also shows the years in which they have committed to achieving carbon neutrality.

Table 1 Net-zero commitments set by major national-level regional level

Status	End target year: national or subnational region	Count
Carbon neutrality achieved	The Republic of Suriname, Bhutan, Comoros, Benin, Gabon, Guyana	6
Carbon neutrality legislation	2030: the Maldives; 2035: Finland; 2040: Austria, Iceland 2045: Germany, Sweden; 2050: Portugal, Fiji, Luxembourg, Ireland, Greece, Australia, Colombia, Switzerland, the United States, the United Kingdom, the European Union, Canada, New Zealand, France, Denmark, Hungary, Japan, South Korea, Chile, Spain; 2070: Nigeria	27
Inclusion of carbon neutrality goals in policies	2030: Dominican Republic, Barbados; 2040: Antigua and Barbuda; 2045: Nepal; 2050: Brazil, Latvia, Cape Verde, Malta, Liberia, Peru, Argentina, Slovakia, Laos, Armenia, Belgium, Singapore, the United Arab Emirates, Vanuatu, Andorra, Uruguay, Croatia, Lithuania, Monaco, Papua New Guinea, Tuvalu, Namibia, Ecuador, the Gambia, Lebanon, Panama, Cambodia, Tunisia, Slovenia, Oman, Georgia, Malaysia, Belize, Ethiopia, Vietnam, Cyprus, Tonga, the Marshall Islands, Italy, Norway, Costa Rica; 2053: Turkey; 2060: Ukraine, China, Saudi Arabia, Russia, Kazakhstan; 2065: Thailand; 2070: India	52
Declaration/pledge of carbon neutrality	2050: South Africa, Estonia, the Federated States of Micronesia, Haiti, Sri Lanka; 2060: Bahrain, Kuwait; 2070: Ghana	8

Data source Oxford Net Zero Tracker (Lang et al. 2023)

Specifically, first of all, **six countries including Bhutan have already achieved carbon neutrality**. Bhutan is the first country in the world to achieve carbon neutrality. It has, since 2009, given priority to environmental policies and committed in its constitution to reserving 60% of its territory as forest land, thus maintaining its innate advantage in forest coverage and enabling it to generate significant forest carbon sinks. Meanwhile, it has also developed hydropower and further offset carbon emissions by exporting this renewable energy.[1]

Secondly, **51 countries such as the United Kingdom and the United States, as well as the European Union, have included carbon neutrality goals in their laws**. The United Kingdom introduced the Climate Change Act in 2008 and amended it in 2019, incorporating the 2050 carbon neutrality goal into it, becoming the first major economy in the world to legislate a commitment to achieving net-zero emissions by 2050. The European Union officially launched the European Green Deal in July 2021, clarifying the goal of reducing greenhouse gas emissions by 55% by 2030, compared to 1990 levels, and planning to achieve carbon neutrality by 2050. As the core policy of the European Green Deal, the EU's "Fit for 55" package of emission reduction measures covers multiple fields such as climate, energy, land use, transportation, and taxation, including new bills such as expanding the EU carbon market, phasing out the sales of fossil fuel vehicles, levying aviation fuel taxes, increasing the proportion of renewable energy, and establishing a carbon border tax. In the case of the United States, after Joe Biden took office as the President of the United States in 2021, he announced the return to the Paris Climate Agreement and put forward the United States' Nationally Determined Contribution goals: reducing greenhouse gas emissions by 50%-52% compared to 2005 by 2030. *The Long-term Strategy of the United States—Pathways to Net-Zero Greenhouse Gas Emissions by 2050* released in November 2021 announced the time nodes and technological pathways for the United States to achieve the ultimate goal of carbon neutrality by 2050.

Thirdly, **52 countries including China and Russia have established national policies for carbon neutrality goals**. China put forward the goal of achieving carbon neutrality before 2060 in September 2020, which has accelerated the global process towards the 2 °C and 1.5 °C temperature increase goals of the Paris Agreement. Russia, on the other hand, has shown its concern for climate issues in the recent years, shifting its stance from being an onlooker to active participation.[2] In 2021, Russia adopted the "Strategy of socio-economic development of the Russian Federation with low greenhouse gas emissions until 2050", stating that Russia will reduce greenhouse gas emissions by 70% by 2030 compared to the 1990 level. According to this strategy, Russia will maintain a social and economic development strategy of low greenhouse gas emissions until 2050, and will achieve carbon neutrality by 2060. The main pathways for Russia to achieve carbon neutrality include reducing the production

[1] TZUNG S. Carbon Negativity In Bhutan: An Inverse Free Rider Problem[R/OL]. Harvard International Review.

[2] Shang Yue, Han Yichen. Meeting the challenges of moving towards a carbon neutral era: Russia's Green New Deal. Modern International Relations, 2021 (10): 18–28.

of fossil fuels, enhancing the structure of low-emission energy, and increasing the carbon absorption capacity of ecosystems such as forests.

Finally, **although eight countries such as South Africa have not issued corresponding laws or policy documents for carbon neutrality plans, they have already made formal commitments to carbon neutrality**. In February 2020, South Africa mentioned in its *Low Emission Development Strategy 2050* that its National Planning Commission was developing a strategy to achieve carbon neutrality by 2050, and that South Africa's low emission development strategy would lay a foundation for its 2050 carbon neutrality goal. In June 2022, the Presidential Climate Commission of South Africa published the *Just Transition Framework* for South Africa, proposing that South Africa should undergo a just transition, aiming to enhance climate resilience and achieve net-zero greenhouse gas emissions by 2050.

The carbon neutrality goals at the main national-level regional level are shown in Table 2. Looking at the content above and Table 2 together, it can be concluded that, although some countries have already established laws and national policies for carbon neutrality goals, the commitments of many countries are still not totally clear. As more and more countries commit to achieving the net-zero emissions goal around the middle of the twenty-first century, the feasibility of achieving the goals of the Paris Agreement is gradually increasing. However, it should also be noted that in order to maintain the feasibility and credibility, it is urgent to transform these commitments into powerful near-term policies and actions and reflect them in the NDCs.

In addition to the carbon neutrality targets at the national-level regional level as described above, sub-national regions such as provinces/states, cities, and urban areas can also set carbon neutrality targets. For example, some states in the United States (such as California, New York, etc.) have put forward the goal of achieving carbon neutrality by the middle of the twenty-first century. Specifically, California in the United States has proposed a carbon neutrality target for 2045, while states like New York, Washington, and Massachusetts in the United States have proposed carbon neutrality targets for 2050. In Australia, states such as New South Wales and Queensland have proposed carbon neutrality targets for 2050. In Japan, regions like Hokkaido and Kanagawa Prefecture have proposed achieving carbon neutrality by 2050. In China, cities like Chengdu, Nanjing, and Qingdao have proposed achieving carbon neutrality by 2050. Shanghai's Chongming District will build a carbon neutrality demonstration area, and Shenzhen's Longgang District will construct a carbon neutrality pilot area.

b. Carbon Neutrality at the Organizational Level

Carbon neutrality at the organizational level refers to the neutralization of greenhouse gas emissions within the organizational boundaries. The specific international standards include ISO 14064-1:2018 *Greenhouse gases—Part 1: Specification with guidance at the organization level for quantification and reporting of greenhouse gas emissions and removals* and ISO/TR 14069:2013 *Greenhouse gases—Quantification and reporting of greenhouse gas emissions for organizations—Guidance for the application of ISO 14064-1*. ISO 14064-1:2018 stipulates the principles and

Table 2 Carbon neutrality targets at the main national-level regional level

Country or national-level region	Carbon neutrality target year	Gases covered	Related documents for carbon neutrality target	Use of offsetting credit outside this region	Interim target
United States of America	2050	All GHGs	Executive order: The Long-term Strategy of the United States—Pathways to Net-Zero Greenhouse Gas Emissions by 2050 Law: Clean Air Act	Not specified	By 2030: reduce GHG emissions by 50–52% below 2005 levels By 2050: net-zero greenhouse gas emissions by mid-twenty-first century
European Union	2050	All GHGs	Government planning: The European Green Deal Law: The European Climate Law, which incorporates climate goals into law	Not specified	By 2030: The European Commission has proposed to increase the current target of at least a 40% reduction in emissions compared to 1990 to at least a 55% reduction (2030 Climate and Energy Framework) By 2050: The European Union as a whole will achieve net-zero emissions covering all sectors and greenhouse gases within its territory and reach climate neutrality
Germany	2050	All GHGs	Law: Climate Action Act Government planning: Climate Action Plan 2050	Not specified	By 2030: The total greenhouse gas emissions will be reduced by at least 55% compared to the baseline level in 1990; the target can be higher but not lower By 2050: greenhouse gas neutrality

(continued)

Table 2 (continued)

Country or national-level region	Carbon neutrality target year	Gases covered	Related documents for carbon neutrality target	Use of offsetting credit outside this region	Interim target
France	2050	All GHGs	Law: Energy and Climate Law No. 1147, 2019 Government planning: Integrated Energy and Climate Plan	No	By 2030: 40% reduction in greenhouse gas emissions compared to 1990 levels (Energy and Climate Law) By 2050: Achieving carbon neutrality
U.K	2050	All GHGs	Law: Climate Change Act	Not specified	By 2030: Economy-wide greenhouse gas emissions reduced by at least 68% compared to 1990 levels (INDC); By 2050: Greenhouse gas emissions reduced by at least 100% compared to 1990 levels
Japan	2050	All GHGs	Policy declaration: Japanese Prime Minister 's Policy Speech Policy planning: Ministry of Economy, Trade and Industry's Green Growth Plan	Not specified	By 2030: GHG emissions reduced by 26% compared to 2013 (or 25.4% compared to 2005) (Japan's new NDC was announced in March 2020, ahead of carbon neutrality) By 2050: Achieving net- zero emissions

(continued)

Table 2 (continued)

Country or national-level region	Carbon neutrality target year	Gases covered	Related documents for carbon neutrality target	Use of offsetting credit outside this region	Interim target
South Korea	2050	Not specified	Policy declaration: The South Korea President announced that the country will achive net-zero emissions by 2050 Submitted to the United Nations: South Korea's 2050 Carbon Neutrality Strategy	Not specified	By 2030: 24.4% reduction in total greenhouse gas emissions compared to 2017 (709.1 $MtCO_2e$) By 2050: Achieving carbon neutrality
Canada	2050	All GHGs	Law: Canada Net Zero Emissions Accountability Act	Yes	By 2030: GHG emissions reduced by 30% from 2005 levels, with economy-wide total emissions of 523 $MtCO_2e$ (unupdated NDC) By 2050: Net zero greenhouse gas emissions
Brazil	2060	Not specified	Submitted NDC	Not specified	By 2025: Reduce greenhouse gas emissions by 37% compared to 2005 levels By 2030: Reduce greenhouse gas emissions by 43% compared to 2005 levels By 2060: Achieving climate neutrality
Argentina	2050	Not specified	Submitted NDC	Not specified	By 2030: Economy-wide emissions limited to 359 $MtCO_2e$ By 2050: Achieving carbon neutrality

(continued)

Table 2 (continued)

Country or national-level region	Carbon neutrality target year	Gases covered	Related documents for carbon neutrality target	Use of offsetting credit outside this region	Interim target
Mexico	2050	Not specified	Policy Declaration	Not specified	By 2030: 36% reduction in greenhouse gas emissions and 51% reduction in black carbon emissions compared to BAU By 2050: Carbon neutrality
South Africa	2050	Not specified	Submitted to the United Nations: Long-term low-emission development strategy	Not specified	2020–2025: Achieve carbon peak, and keep emissions stable for ten years after the peak By 2050: Absolute emissions drop to 212–428 $MtCO_2e$ By 2050: Achieving net zero emissions

requirements for quantifying and reporting greenhouse gas emissions and removals at the organizational level, including the requirements for the design, development, management, reporting and verification of an organization's greenhouse gas inventory. According to the definition of ISO, an organization can be a commercial one like sole-trader, company, enterprise, and firm, or authority, institution, charity, and association, whether incorporated or not, public or private.

Based on the statistics of the University of Oxford as of the end of October 2023, 900 companies worldwide have already announced their carbon neutrality targets. Among them, 36 companies have been externally verified as carbon neutral or have themselves declared that they have achieved carbon neutrality. For example, the German automotive supplier Bosch announced that it achieved carbon neutrality in 2020. L'Oréal will achieve carbon neutrality in all its factories by 2025. Unilever, IKEA and others will achieve carbon neutrality by 2030. Mittal, Shell, BP and others will achieve carbon neutrality by 2050. The carbon neutrality targets of typical companies in China and abroad are shown in Table 3.

For companies, announcing carbon neutrality targets brings many benefits. On the one hand, the company can directly benefit from carbon neutrality, such as reducing the cost of future carbon reduction through immediate action, enhancing brand and product differentiation to win new customers and increase brand loyalty from existing customers, improving brand awareness and positioning products in major markets, etc. On the other hand, companies can also benefit indirectly from carbon neutrality,

Table 3 Carbon neutrality targets of typical domestic and foreign companies

Enterprise	Carbon neutrality goal
Baidu	Achieve carbon neutrality at the operational level of the Group by 2030
Tencent	Achieve full carbon neutrality in its operations and supply chain no later than 2030
Alibaba	Achieve carbon neutrality in its operations by 2030
Geely	Achieve full-chain carbon neutrality by 2045
Bosch	Achieve carbon neutrality by 2020
Unilever	Achieve carbon neutrality by 2030
IKEA	Achieve carbon neutrality by 2030
Amazon	Half of transport to be zero carbon by 2030
L'Oreal	All factories to be carbon neutral by 2025
Schneider	Zero-net emissions at the operational level by 2030 and zero net emissions in the supply chain by 2050
Benz	Achieve carbon neutrality for vehicles by 2039
Volkswagen	Achieve zero-emission vehicles by 2050
Danone	Carbon neutrality across the entire industry chain by 2050
Nestle	Net-zero carbon emissions by 2050
Shell	Net-zero carbon emissions by 2050
PetroChina	Net-zero carbon emissions by 2050

such as establishing a reputation in climate and sustainable development, attracting and retaining employees, etc. More importantly, the carbon neutrality of companies can reduce global carbon emissions, which is of great significance for the global climate response.

The carbon neutrality efforts of companies in China started relatively late. After China announced its carbon neutrality target, the companies that took the lead in launching carbon neutrality plans were mostly state owned, such as China National Nuclear Corporation, China National Petroleum Corporation, China National Offshore Oil Corporation, China Energy Investment Corporation, China Academy of Building Research, and State Power Investment Corporation. Other Chinese companies have also successively announced or are currently planning carbon neutrality actions. For example, Tongwei Group planned to achieve corporate carbon neutrality in 2023. Ant Financial aimed to achieve carbon neutrality in its operations in 2021 and net-zero emissions by 2030. China Three Gorges Corporation planned to achieve corporate carbon neutrality in 2040. At the "Corporate Day of China Pavilion" during the 27th Conference of the Parties to the UNFCCC (COP27) in November 2022, many companies presented their carbon emission reduction commitments. Alibaba introduced the carbon reduction target of "Scope 3+" for its platform ecosystem, promising to achieve carbon neutrality in its own operations no later than 2030 and to halve the carbon emission intensity of the upstream and downstream value chain no later than 2030. Vanke stated that it planned to install solar photovoltaic power generation systems in at least 18 shopping malls by 2025. Newlink Naas, the first listed company in China's charging service sector, introduced that it had achieved a carbon emission reduction of 700,000 tonnes CO_2e in the first half of 2022, which was nearly 80% of the total carbon emission reduction in 2021. During China's green opportunity period of "dual carbon" goals, numerous Chinese companies have already taken actions to jointly create a zero-carbon future.

c. Carbon Neutrality at the Project Level

Carbon neutrality at the project level refers to the neutralization of greenhouse gas emissions throughout the entire life cycle of the project (the entire industrial chain), which involves carbon neutrality across the entire industrial chain.

In addition, event organizers can also set carbon neutrality goals for large-scale events. There have been many typical cases of carbon neutral events around the world, such as the 2006 Turin Winter Olympics, the 2008 Beijing Olympics, the 2010 South Africa World Cup, the 2016 G20 Hangzhou Summit, the Beijing 2022 Winter Olympics, etc. Among them, the Turin Winter Olympics is the first Olympic event to achieve "carbon neutrality". The organizing committee launched the "Turin Climate Heritage" activity and carried out carbon offsetting through forestry, energy conservation and emission reduction and renewable energy projects; The World Cup in South Africa has been declared carbon neutral through the purchase of assessed and qualified carbon offset sources. In March 2022, the Technology and Industrialization Development Center of the Ministry of Housing and Urban–Rural Development of China selected a number of projects as typical examplary cases of carbon peaking and carbon neutrality in the urban and rural construction field, including zero-carbon parks, near-zero carbon buildings, and green and low-carbon renovation projects, low-carbon operation cases and zero-carbon communities, such as the 10th China Flower Expo Park in Chongming, Shanghai, the Demonstration Building of the China Academy of Building Sciences, the Demonstration Center of Shenyang Jianshe University, and the Central South Office Building of Hubei Institute of Construction Research.

d. Carbon Neutrality at the Product Level

Carbon neutrality at the product level refers to achieving carbon neutrality throughout the entire life cycle of a product. The specific international standard is ISO 14067:2018 *Greenhouse gases—Carbon footprint of products—Requirements and guidelines for quantification.* ISO 14067:2018 defines the principles, requirements and guidelines for quantifying carbon footprint at the product level. Its goal is to quantify greenhouse gas emissions associated with the life cycle of a product, beginning with resource extraction and raw material sourcing and extending through the production, use and end-of-life stages of products.

Some companies set carbon neutrality targets for all or some of their products. A small number of companies set carbon neutrality targets for all of their products, with Apple being a representative example. Apple said it has achieved carbon neutrality in its global operations and plans to achieve carbon neutrality for its entire business, production supply chain and product life cycle by 2030. Most companies only plan to achieve carbon neutrality for individual products, such as Mercedes-Benz and Nestle.

e. Differences and Connections Among Carbon Neutrality at the Regional, Organizational, Project and Product Levels

The carbon neutrality goals at the four levels of region, organization, project, and product complement and promote each other, and their common goal is to respond to global climate change. In the process of responding to global climate change, carbon neutrality goals at the regional level alone will be isolated and could not get help, and carbon neutrality goals at the organizational, project or product levels alone will be a drop in the bucket. It is necessary to contribute to reducing global carbon dioxide emissions at the regional, organizational, project and product levels. The carbon neutrality actions at the four levels should share the same goal and are consistent, mutually reinforcing, and none of them can be missing.

From the perspective of the timeline of carbon neutrality, carbon neutrality actions at the regional, organizational, project and product levels can be carried out simultaneously or in sequence. On the one hand, while countries and regions are setting overall carbon neutrality goals, organizations, projects and products can start from themselves and contribute from the bottom level up to the implementation of regional carbon neutrality goals. On the other hand, countries and regions can first establish carbon neutrality goals at the regional level and encourage organizations and project implementation units to define their own carbon neutrality goals; organizations, projects and products can also take the lead in formulating carbon neutrality goals to help with the formulation of regional carbon neutrality goals.

From the perspective of the forms and purposes of carbon neutrality, there are certain differences in the carbon neutrality actions at the regional, organizational, project and product levels. Carbon neutrality at the regional level is generally proposed by national or regional governments and pledged in the form of legal documents, policy planning, and NDC targets. Its purpose is to contribute to global sustainable development, show the image of responsible countries or regions, and serve as a specific form of international climate change agreements such as the Paris Agreement and NDC targets. Carbon neutrality targets at the regional level have high credibility and strong effectiveness in execution, and can drive a series of regions and industries to take carbon neutrality actions. The scale of planning impact and the difficulties of policy implementation are both very large. Carbon neutrality at the organizational level is generally put forward by enterprises in their corporate environmental progress or social responsibility reports, with the purposes of increasing brand awareness, assuming social responsibilities and increasing corporate profits. Its implementation is rather difficult. Carbon neutrality at the project and product levels is proposed by product manufacturers or activity initiators, and usually achieved through offsetting methods. Its goals are to increase product competitiveness, attract consumers and guide them to consume in a low-carbon style, and promote the whole industrial chain of products to reduce emissions. The implementation of project carbon neutrality is less difficult and has a shorter duration.

From the perspective of carbon accounting methods, carbon accounting at the regional, organizational, project and product levels has its own standards and significance, and cannot be simply aggregated. Carbon accounting at the regional level reflects greenhouse gas emissions as accurately as possible from a physical perspective. Countries or governments usually compile national greenhouse gas emission inventories and make specific arrangements for emission classification, inventory standards and emission lists. The purpose of carbon accounting at the organizational level is to quantify the attribution of carbon emission responsibilities, while the significance of carbon accounting at the product level is to guide consumers to consume in a low-carbon style. Although carbon accounting at the regional, organizational and project levels is usually expressed in terms of carbon dioxide equivalent, their significance is different and they cannot be simply aggregated. If integration is to be done, issues such as double counting need to be taken into account.

2 Accurately Understanding the Carbon Neutrality Goals

2.1 Concepts Involved in Carbon Neutrality

To understand the carbon neutrality goals more accurately, this section will specifically introduce the concepts that frequently appear in carbon neutrality. On the one hand, it will sort out the differences and connections among similar concepts such as carbon neutrality, net-zero carbon emissions, net-zero emissions, and climate neutrality; on the other hand, it will introduce concepts related to carbon neutrality such as carbon peaking and carbon footprint.

a. Carbon Neutrality

The Special Report on Global Warming of 1.5 °C released by the IPCC points out that Carbon Neutral, or Carbon Neutrality means that within a certain period, anthropogenic carbon dioxide emissions and anthropogenic carbon dioxide removals reach a balance on a global scale. In that definition, anthropogenic carbon dioxide emissions refer to carbon dioxide emissions brought by human activities, including emissions from the combustion of fossil fuels, industrial processes, agricultural and land use activities, etc.; anthropogenic carbon dioxide removals refer to the removal of carbon dioxide from the atmosphere by humans, including increasing carbon absorption through afforestation, carbon capture and storage, etc.

In short, carbon neutrality is the balance between reducing emission sources and increasing carbon sinks (i.e. removal and absorption). Enterprises, organizations or individuals calculate the total amount of greenhouse gas emissions directly or indirectly generated within a certain period of time (usually one year), and then offset their own carbon dioxide emissions through forms such as afforestation and energy conservation and emission reduction to achieve carbon neutrality.

However, some literature believes that carbon neutrality includes not only carbon dioxide but also all the other greenhouse gases. For example, carbon neutrality in the PAS 2060, the specification guide for carbon neutrality, refers to all greenhouse gases. Therefore, whether it is a country, an organization or a project, when proposing carbon neutrality, it should clearly explain its scope.

b. Net-Zero Carbon Emissions and Net-Zero Emissions

According to the definition of the IPCC, net-zero carbon emissions (Net-Zero Carbon Emissions) is the short form of net-zero carbon dioxide emissions (Net-Zero CO_2 Emissions), and its meaning is the same as that of "carbon neutrality". Net-zero emissions (Net-Zero Emissions) means that within a certain period, the emissions of all greenhouse gases (GHGs) (measured in carbon dioxide equivalent) caused by human activities of an organization in one year and the absorption of anthropogenic emissions reach a balance.

c. Climate Neutrality

According to the definition of the IPCC, climate neutrality refers to a state in which human activities have no net impact on the climate system. To achieve this state, it is necessary to balance residual carbon emissions and carbon removals while considering the regional or local biogeophysical effects of human activities.

d. Carbon Peaking

Carbon peaking refers to the process in which the annual carbon dioxide emissions in a certain region or industry reach the highest historical value and no longer increase, and then enter a continuous decline after experiencing a plateau period. Carbon peaking is the historical turning point at which carbon dioxide emissions change from increasing to decreasing, marking the decoupling of carbon emissions from economic development. The peaking goal includes the peaking year and the peak value.

e. Carbon Footprint

Carbon footprint refers to the collection of greenhouse gas emissions caused by corporate organizations, activities, products or individuals through transportation, production and consumption, and various production processes. The general accounting includes emissions from the entire life cycle, including raw material processing and transportation, use, and waste disposal.

2.2 Core Elements of Carbon Neutrality

After understanding the core concept of carbon neutrality, this section will introduce the core elements of carbon neutrality. First, it will introduce at what levels the carbon neutrality target can be built, what time frames are generally set for carbon neutrality at different levels, what types of greenhouse gases, emission types, and covered industries are covered by the carbon neutrality target, how to calculate carbon emissions to determine whether the carbon neutrality target has been achieved, and what are the ways to commit to the carbon neutrality target. Secondly, in addition to carbon emission reduction and carbon absorption, ways to achieve carbon neutrality goals also include offsetting, which will also be discussed in this section.

a. Levels of Carbon Neutrality

According to the differences in the subjects of carbon neutrality, carbon neutrality goals can be categorized into four levels: regional, organizational, project, and product. There has been a brief discussion in the first section of this chapter, and Chaps. 3, 4, 5, and 6 of this book will respectively discuss in detail the implementation pathways of carbon neutrality at these four levels.

b. Time Frame for Carbon Neutrality

Regions and enterprises usually set carbon neutrality targets and phased targets. Neutrality targets at the regional and organizational levels usually include a specific time point, such as achieving carbon neutrality by 2050 or 2060, and lower-level targets will also be proposed. For example, Shell's long-term goal is to achieve net-zero emissions by 2050 or earlier. To this end, Shell has set decomposition targets: by 2035, the carbon intensity of energy products sold will be reduced by 30% compared to 2016, and by 2050 by 65%; the target for 2021 is to reduce carbon intensity by 3% to 4% compared to 2016. At the project and product level, the time frame for carbon neutrality is usually the entire life cycle of the project or product.

c. Coverage of Carbon Neutrality

The coverage of carbon neutrality is the key point of carbon neutrality goals, including the types of greenhouse gases in the scope of the carbon neutrality goals, whether it includes direct emissions or indirect emissions, as well as the covered industry fields and whether it includes international aviation and shipping, etc.

(i) Types of Greenhouse Gases Included

When setting a carbon neutrality target, it is necessary to determine the types of greenhouse gases included, whether it includes only carbon dioxide or all greenhouse gases (GHGs) (including carbon dioxide, methane, nitrous oxide, hydrofluorocarbons, perfluorocarbons, sulfur hexafluoride and nitrogen trifluoride). The carbon neutrality targets of some countries or regions include all GHGs, such as the European Union, Germany, the United Kingdom, France, Italy, etc. In particular, New Zealand's carbon neutrality target includes all GHGs except biogenic methane emitted from animal and plant sources. The carbon neutrality targets of other countries clearly do not include all GHGs, such as South Africa, New Zealand, Rwanda, etc.

(ii) Direct and Indirect Emissions

Carbon emissions are usually classified into three types: Scope 1 emissions are direct emissions, Scope 2 emissions are energy indirect emissions, and Scope 3 emissions are other indirect emissions, as shown in Table 4.

Table 4 Emission types in carbon neutrality goals

Emission type	Description	Notes
Scope 1	Direct emissions	Greenhouse gases emitted directly into the atmosphere within the physical boundaries or controlled assets of an enterprise, including emissions from stationary combustion sources, mobile combustion sources, direct emissions from industrial process, fugitive emissions from anthropogenic systems, etc
Scope 2	Energy indirect emissions	Indirect emissions caused by the use of externally supplied electricity, heat, steam, compressed air, etc. of an enterprise
Scope 3	Other indirect emissions	All other indirect emissions generated by an enterprise's production and operation, including emissions from transportation (freight transportation, employee commuting, business travel, etc.), emissions from products used (raw materials, equipment, buildings, etc.) or services (consulting, cleaning, maintenance, etc.), emissions from the use and end-of-life treatment of products produced, etc

Scope 1 emissions refer to greenhouse gases emitted directly into the atmosphere within the physical boundaries of an enterprise or the assets it controls, including emissions from the combustion of stationary sources, emissions from the combustion of mobile sources, emissions directly from industrial processes, and fugitive emissions from man-made systems.

Scope 2 emissions refer to the indirect emissions caused by an enterprise due to the use of externally supplied electricity, heat, steam, compressed air, etc. Regarding the calculation of the grid emission factor, the World Resources Institute (WRI) first introduced the concepts of location-based grid electricity emissions and market-based grid electricity emissions in a guide on calculating carbon emissions from electricity use (GHG Protocol Scope 2 Guidance) published in 2014. First, location-based grid electricity emissions. When calculating location-based grid electricity emissions, an enterprise's electricity emissions are calculated based on the average emission factor of the entire grid's emissions, without considering whether the enterprise uses clean electricity or fossil electricity. Second, market-based grid electricity emissions. When calculating market-based grid electricity emissions, the electricity emission factors provided by electricity traders to each enterprise are used for calculation. If the electricity provided by the electricity trader to the enterprise is clean electricity and the ownership of its clean attributes can be proved, it can be considered that the enterprise's electricity emissions are zero. Third, the use of clean electricity. The following ways can all claim to use clean electricity: direct supply from an enterprise's own power station and declaring not to claim environmental attributes; non-direct supply from an enterprise's own power station, claiming environmental attributes but for self-use; direct purchase of clean electricity and being able to prove the environmental attribute ownership of clean electricity; purchasing electricity from the grid and purchasing an equal amount of environmental rights and interests (such as green certificates).

Scope 3 emissions refer to all other indirect emissions except Scope 2. The latest ISO 14064-1:2018 has classified Scope 3 emissions, including emissions from transportation (goods transportation, employee commuting, business travel, etc.), emissions from the use of products (raw materials, equipment, buildings, etc.) or services (consulting, cleaning, maintenance, etc.), and emissions generated in the use and end-of-life stages of produced products, etc. Due to the large amount of external data involved in Scope 3 emissions, the difficulty of accounting and management is relatively large. Except for some excellent enterprises that have considered Scope 3 emissions in carbon accounting and setting emission reduction goals, most enterprises do not consider Scope 3 emissions. In particular, the Science Based Targets initiative (SBTi) requires that if Scope 3 emissions exceed 40% of the total emissions, Scope 3 emissions must be considered when setting emission reduction goals.

At the regional and project levels, carbon neutrality targets generally do not involve direct and indirect emissions. For carbon neutrality targets at the national level, since it is a calculation of the overall carbon emissions of a region, carbon neutrality targets generally do not involve the question of direct and indirect emissions, or rather, they are all considered as direct emissions. However, at the provincial and city levels, it is necessary to distinguish between direct and indirect emissions, such as large-scale cross-regional power transmission, which will lead to disputes over carbon attribution.

At the organizational level, it is necessary to distinguish between direct emissions and indirect emissions. The calculation of direct and indirect emissions has a great impact on the power of the carbon neutrality target. If the distinction is not made, the carbon neutrality target itself will become ambiguous.

At the product level, emissions are generally calculated based on the entire life cycle of the product, i.e. the carbon footprint of the product. For example, the production of a mobile phone involves the carbon emissions generated during the entire process of manufacturing all parts of the mobile phone, from raw material mining to the disposal of the mobile phone.

When setting carbon neutrality goals, companies must first determine the boundaries of carbon neutrality coverage, whether it is Scope 1 emissions, Scope 1 + Scope 2 emissions, Scope 1 + Scope 2 + Scope 3 emissions, or carbon emissions from the supply chain and products.

(iii) Does Regional Carbon Neutrality Include International Aviation and Shipping?

The industries covered by the carbon neutrality targets of different actors will also be different. In particular, the international aviation and shipping industries will involve the jurisdiction of multiple countries, and there are certain difficulties in the supervision, reporting, and verification of carbon emissions. The international community has not reached a consensus on this. Therefore, different countries have different carbon neutrality targets for whether to include international aviation and shipping. Some countries have made it clear that their carbon neutrality targets include international aviation and shipping (with great efforts), including the United Kingdom, Austria, Fiji, etc. Other countries have clearly excluded international aviation and shipping (with less efforts), such as Germany, France, the Netherlands, Sweden, etc.

d. Carbon Emissions Accounting Methods and Standards

(i) Regional Level

For the carbon neutrality goal at the regional level, the government usually formulates a national greenhouse gas inventory, which is an important basis for judging the total amount of domestic carbon dioxide control and the basis for international climate negotiations. China compiled a national greenhouse gas inventory in 1994, 2005, 2012, 2014 and 2017, and each province and city also compiled a provincial greenhouse gas inventory and a city and county greenhouse gas inventory.

The accounting method of regional carbon emissions is mainly based on the *Guidelines for National Greenhouse Gas Inventories* issued by the IPCC. It is mainly divided into greenhouse gas emissions from energy, industrial processes and product use, agriculture, forestry and other land use, and waste.

(ii) Organizational Level

The carbon emission accounting method at the organizational level originated from the Greenhouse Gas Protocol issued by WRI and the subsequent ISO 14064-1: 2018. In China, the National Development and Reform Commission and the Ministry of Ecology and Environment issued the *Corporate Greenhouse Gas Emissions Accounting Method and Reporting Guide* for power generation companies, power grid companies, steel production companies, chemical companies and other industries. The reference document for the accounting method of carbon accounting is GHG protocol, and emission factors can refer to IPCC 2006, national industry greenhouse gas accounting and reporting guidelines, national average power grid emission factors, etc.

(iii) Project and Product Level

The carbon emission accounting method at the project and product level is mainly based on the emission of the product from cradle to grave (LCA). The carbon neutrality guidance document at the project and product level is ISO 14067, and the carbon emission accounting method reference document of WRI's *Product Life Cycle Accounting and Reporting Standard is BSI's PAS 2050—Specification for the assessment of the life cycle greenhouse gas emissions of goods and services. PAS 2050—Specification for the assessment of the life cycle greenhouse gas emissions of goods and services* is the world's first product carbon footprint method standard. It was published by the British Standards Institution in October 2008 and is widely used by companies to evaluate the greenhouse gas emissions of their goods and services. The emission factor reference documents at the product level are the China Life Cycle Database (CLCD), the United States Life Cycle Database (U.S. LCD), the Ecoinvent database, etc.

e. Carbon Neutrality Offsetting Methods

There are three main ways to achieve carbon neutrality: direct emission reduction, carbon offsetting, and carbon sink. Carbon offsetting refers to the purchase of environmental rights to offset the remaining emissions. Since allowing the use of carbon offsets will damage the effect of carbon neutrality actions, there has been controversy over whether carbon offsets are allowed in carbon neutrality.

At the regional level, the determination method of carbon neutrality is very clear. That is, if the greenhouse gas emissions generated within the region are equal to or lower than the absorption amount, then carbon neutrality can be deemed as achieved. Some countries or regions have relatively ambitious carbon neutrality targets and clearly state that their carbon neutrality targets do not include international offsets. Their unabated emissions can be absorbed and eliminated through afforestation, carbon capture and storage (CCS) technologies, such as the European Union, Switzerland, Portugal, Iceland, Fiji, etc. Some countries have less ambitious carbon neutrality targets and clearly state that their carbon neutrality targets include international offsets, such as the United Kingdom, Sweden, Norway, Ireland, New Zealand, etc. And most regions have not yet specified whether international offsets are allowed for the time being.

At the organizational level, companies announce whether to use offsets to achieve carbon neutrality targets based on their own situations. Since it is rather difficult for organizations to absorb carbon emissions through afforestation within limited physical boundaries, most companies will use offsetting indicators to achieve carbon neutrality targets.

At the project level, due to the geographical and time limitations of the activities themselves, carbon offsetting is mostly needed to neutralize the unabated carbon dioxide emissions.

f. Ways of Committing to Carbon Neutrality

For regions, the contents of commitment to carbon neutrality include carbon neutrality in a narrow sense, net-zero emissions or climate neutrality, and the ways of committing are classified as legislation, inclusion in policies, pledges or commitments, etc. According to the carbon neutrality tracking dataset published by the University of Oxford, as of the end of October 2023, 93 countries and 339 subnational regions (provinces, states, cities, etc.) have formally committed to carbon neutrality (or net-zero emissions, climate neutrality) targets. Among them, six countries such as Suriname and Bhutan and two subnational regions such as Sikkim in India have already achieved carbon neutrality targets; 27 countries such as the United Kingdom, France, Switzerland, the European Union, Canada, South Korea, and 31 subnational regions such as California in the United States and Nova Scotia in Canada have formulated carbon neutrality targets in the form of legislation; 52 countries such as China, the United States, Finland and 168 subnational regions such as Nouvelle-Aquitaine in France have made commitments in the form of policy documents; and eight countries such as South Africa and 138 subnational regions such as Ulsan in South Korea have not yet issued laws or policy documents specifically for carbon neutrality but have already made formal commitments.

For organizations, carbon neutrality targets are distinguished as carbon neutrality in a narrow sense, net-zero emissions, climate neutrality, 100% renewable energy, science-based targets (SBT), etc., and the ways of committing include writing into organizational strategies, commitments or announcements, etc. At the organizational level, companies and enterprises are important players in achieving carbon neutrality targets. The carbon neutrality tracking dataset published by the University of Oxford has collected the progress of companies worldwide in carbon neutrality targets. As of the end of October 2023, a total of 900 companies have made commitments or progress in carbon neutrality targets, including five companies such as Netflix that have been externally verified to have achieved carbon neutrality, 31 companies such as Tokio Marine Holdings that have self-declared to have achieved carbon neutrality, 672 companies such as Apple that have written carbon neutrality targets into their corporate strategies, and 192 companies such as Air Canada that have not reflected them in their corporate strategies but have already made commitments.

At the organizational level, the ways to achieve carbon neutrality targets can be classified as absolute emission reduction and relative emission reduction. On the one hand, absolute emission reduction refers to the rigid emission reduction that is not offset by any environmental rights and interests such as emission reduction rights, green certificates, and green electricity. This kind of emission reduction can often only be achieved through energy conservation and efficiency improvement and fuel substitution. The goal of absolute emission reduction is generally that the absolute emissions in the target year are reduced by a certain fixed percentage compared to the base year. On the other hand, the way of relative emission reduction is much more flexible and also includes the use of environmental rights and interests. When companies propose relative emission reduction targets, they can simultaneously put forward multiple phased targets, such as the proportion of clean electricity used in the target year, the reduction ratio of emissions per product compared to the baseline year, etc. In relative emission reduction, companies can purchase a variety of environmental rights and interests to offset their own emissions, such as REC, IREC, TIGRs for clean electricity, CER, CCER, VER, GSCER, GSVER, REDD, REDD + , CCBS for emission reduction rights. The highly recommended one is the Nature-based Solution NbS), that is, to slow down climate change through the protection, restoration and sustainable management of ecosystems such as afforestation. Companies can also achieve sustainable development goals by formulating a most suitable combination of renewable energy tools (green tariff, EAC (Energy Attribute Certificates), EAC plus, power purchase agreements (PPA) and on-site renewable energy (on-site RE)).

3 China's Practices in Carbon Peaking and Carbon Neutrality

3.1 Evolution of China's Carbon Reduction Targets

China's carbon reduction targets mainly go through three stages: the greenhouse gas emission intensity target set at the Copenhagen Climate Conference in 2009, the greenhouse gas emission peaking target set at the Paris Climate Conference in

2015, and the carbon peaking and carbon neutrality targets set at the United Nations General Assembly in 2020.

In the first stage, at the World Climate Conference in Copenhagen in December 2009, China proposed that by 2020, China's carbon dioxide emissions per unit of GDP would be 40–45% lower than those in 2005. In 2018, China achieved this goal ahead of schedule.

In the second stage, at the Paris Climate Conference in December 2015, President Xi Jinping pledged that China's carbon dioxide emissions would peak around 2030 and strive to achieve it as early as possible.

In the third stage, on September 22, 2020, President Xi Jinping delivered an important speech at the general debate of the 75th United Nations General Assembly, stating that China would increase its nationally determined contributions, adopt more powerful policies and measures, strive to amake carbon dioxide emissions peak before 2030, and work hard to achieve carbon neutrality before 2060 (hereinafter referred to as the China's Dual Carbon Targets). This was also the first time that China clearly stated the timeline for achieving carbon neutrality. In December 2020, China also announced that it would increase its nationally determined contributions. By 2030, China's carbon dioxide emissions per unit of GDP would be reduced by more than 65% compared to 2005, the proportion of non-fossil energy in primary energy consumption would reach about 25%, the forest stock volume would increase by 6 billion cubic meters compared to 2005, and the total installed capacity of wind power and solar power would reach more than 1.2 billion kilowatts. China's 14th Five-Year Plan mandates the formulation of action plans for the peaking of carbon dioxide emissions before 2030 and adopt stronger policy measures to achieve carbon neutrality before 2060.

Internationally, President Xi Jinping's proposal of China's "30·60 Goals" reflects the country's high attention to "carbon neutrality" and its attitude and determination to firmly fulfill the goal of "carbon neutrality". On September 30, 2020, President Xi Jinping's speech at the United Nations Summit on Biodiversity showed that China would actively participate in global environmental governance, earnestly fulfill its obligations under environmental-related treaties such as those on climate change and biodiversity, increase its nationally determined contributions, and adopt more powerful policies and measures to achieve the "30·60 Goals". On November 12, 2020, in his speech at the Paris Peace Forum 2020 (3rd edition), President Xi Jinping pointed out that China would adhere to the concept of green development, be committed to implementing the Paris Agreement on climate change, and formulate and implement plans for the "30·60 Goals". On November 17, 2020, in his speech at the 12th BRICS Summit, President Xi Jinping stated that China would abide by the principle of common but differentiated responsibilities and would fulfill its promise regarding the "30·60 Goals". On December 12, 2020, in his speech at the Climate Ambition Summit, President Xi Jinping reiterated the "30·60 Goals" and was willing to further announce that by 2030, China's carbon dioxide emissions per unit of GDP would be reduced by more than 65% compared to 2005, the proportion of non-fossil energy in primary energy consumption would reach about 25%, the forest stock volume would increase by 6 billion cubic meters compared to 2005, and the total installed capacity of wind power and solar power would reach more than 1.2 billion kilowatts.

Domestically, the work on "carbon peaking and carbon neutrality" has been placed in an important position in the government's work. At the Central Economic Work Conference on December 18, 2020, "doing a good job in carbon peaking and carbon neutrality" was listed alongside strengthening the country's strategic scientific and technological strength, expanding domestic demand, and ensuring food security, becoming one of the eight key tasks of the country in 2021, reflecting the country's high attention to "carbon neutrality" and its attitude and determination to incorporate the goal of "carbon neutrality" into the implementation stage. The Central Economic Work Conference pointed out that it is necessary to promptly formulate action plans for the peaking of carbon emissions before 2030 and support localities with conditions to take the lead in the peaking. It is necessary to accelerate the adjustment and optimization of the industrial structure and energy structure, promote the early peaking of coal consumption, vigorously develop new energy, accelerate the construction of the national energy use rights and carbon emission rights trading markets, and improve the dual control mechanism (energy intensity and total consumption). It is necessary to continue to fight the tough battle against pollution and achieve the synergy of pollution reduction and carbon reduction. It is necessary to undertake large-scale land greening actions and enhance the carbon sink capacity of the ecosystem. At the 9th meeting of the Central Committee for Financial and Economic Affairs on March 15, 2021, President Xi Jinping pointed out, "Achieving carbon peaking and carbon neutrality is a broad and profound systemic transformation of the economy and society. We should incorporate carbon peaking and carbon neutrality into the overall layout of ecological civilization construction, work with determination, and achieve the goals of carbon peaking before 2030 and carbon neutrality before 2060 as scheduled." In October 2022, President Xi Jinping clearly defined the overall goals for China's development by 2035 in the report of the 20th National Congress of the Communist Party of China, one of which is "Widely form a green production and lifestyle, with carbon emissions steadily declining after peaking, the ecological environment fundamentally improved, and the goal of a beautiful China basically achieved." The report of the 20th National Congress of the Communist Party of China also made arrangements for "actively yet steadily promoting carbon peaking and carbon neutrality." At the end of October 2021, the State Council Information Office published the White Paper on China's Policies and Actions for Addressing Climate Change, stating that China has overcome its own economic, social and other difficulties as a developing country, actively assumed the responsibility for international climate governance, introduced and implemented a series of strategies, measures and actions to address climate change, firmly followed the pathway of green development, and achieved positive results in low-carbon emission reduction.[3]

[3] Information Office of the State Council of the People's Republic of China. White Paper on China's Policies and Actions to Address Climate Change [EB/OL]. Information Office of the State Council of the People's Republic of China (2021-10-27) [2023-11-06].

3.2 *China's 1 + N Policy Framework for Dual Carbon Goals*

In recent years, China has implemented a national strategy to actively respond to climate change, continuously strengthened its goals for independent contributions, and accelerated the construction of the "1 + N" policy system for dual carbon goals. On the one hand, the "1" in "1 + N" refers to the guiding opinions on carbon peaking and carbon neutrality. On October 24, 2021, the Communist Party of China Central Committee and the State Council issued the "Guiding Opinions on Comprehensive Implementation of New Development Philosophy for Dual Carbon Targets (CPC Central Committee Document [2021] No.33)" (hereinafter referred to as the "Guiding Opinions for Dual Carbon Targets"). It put forward the overall goals at the national level for 2025, 2030 and 2060, and pointed out eight key strategies and three important safeguard measures. On October 26, 2021, the State Council released the "Action Plan for Carbon Peaking Before 2030", which further clarified the main goals at the national level during the two important transformation periods of the "14th Five-Year Plan" and the "15th Five-Year Plan", and clearly stated 10 key tasks and 43 subdivided task areas. The "Guiding Opinions for Dual Carbon Targets'" and the "Action Plan for Carbon Peaking Before 2030" have clarified the timetable and roadmap for the work of carbon peaking and carbon neutrality. Together, they serve as the "1" in the "1 + N" policy system and play a leading role in the work of "carbon peaking and carbon neutrality".

On the other hand, the "N" in the "1 + N" policy system refers to a series of documents issued according to different fields and industries related to "carbon peaking and carbon neutrality". In June 2022, 17 departments including the Ministry of Ecology and Environment, the National Development and Reform Commission, and the Ministry of Science and Technology jointly issued the "National Climate Change Adaptation Strategy 2035", making overall planning and deployment for the work of adapting to climate change from the present to 2035.

Also in June 2022, seven departments including the Ministry of Ecology and Environment, the National Development and Reform Commission, and the Ministry of Industry and Information Technology jointly issued the "Implementation Plan for Synergistic Pollution and Carbon Reduction", making systematic arrangements for promoting the synergy of pollution reduction and carbon reduction.

"N" involves the action plan for carbon peaking before 2030, as well as policy measures and actions in key fields and industries. Mr Xie Zhenhua, China's Special Envoy for Climate Change Affairs, pointed out that "N" specifically includes ten aspects such as optimizing the energy structure, promoting the optimization and upgrading of industries and industry, advancing energy-efficient and low-carbon buildings and low-carbon infrastructure, building a green and low-carbon transportation system, developing a circular economy, promoting innovation in green and low-carbon technologies, developing green finance to expand financial support and investment, introducing supporting economic policies and reform measures, establishing and improving the carbon trading market, and implementing nature-based solutions. The "White Paper on China's Policies and Actions for Addressing Climate Change" pointed out that "N" includes not only the implementation plans for sub-fields and sub-industries such as national energy, industry, urban and rural construction, transportation, and agriculture and rural areas, but also the safeguard plans for science and technology, fiscal system, finance, pricing, carbon sinks, energy transformation, and synergy of pollution reduction and carbon reduction that support these implementation plans. In addition, all 31 provinces, autonomous regions, and municipalities directly under the Central Government have formulated their own implementation plans for carbon peaking in their respective regions. The above series of documents will build a carbon peaking and carbon neutrality policy system with clear goals, reasonable division of labor, effective measures, and orderly connection.

(a) Key Areas of China's "Dual Carbon" Goals

The State Council's "Action Plan for Carbon Peaking Before 2030" proposes to implement the "Ten Actions for Carbon Peaking", aiming to achieve the "carbon peaking and carbon neutrality" goals in ten key areas. Specifically, these include actions for the green and low-carbon energy transition, actions for energy conservation, carbon emissions reduction and efficiency improvement, actions for carbon emissions peaking in industry, actions for carbon peaking in urban and rural development, actions for the green and low-carbon transport, actions for carbon emissions reduction through the circular economy, actions for encouraging innovation in green and low-carbon science and technology, actions for consolidating and enhancing carbon sink capacity, and actions for green and low-carbon lifestyle nationwide. The "White Paper on China's Policies and Actions for Addressing Climate Change" points out that China will successively release implementation plans for carbon peaking in key areas and a series of supporting and safeguarding measures; and will promote actions to adapt to climate change in key areas such as agriculture, forestry and grasslands, water resources, and public health. Specifically, the relevant policy plans in key areas under China's "Dual Carbon" goals are shown in Table 5.

(b) Key Industries of China's "Dual Carbon" Goals

The key industries covered by China's carbon peaking and carbon neutrality policies include coal, oil and natural gas, steel, building materials, petrochemicals and chemicals, non-ferrous metals, light industry, textiles, hydrogen energy, renewable energy and so on. The "White Paper on China's Policies and Actions for Addressing Climate Change" points out that during the "13th Five-Year Plan" period, the transformation and upgrading of key industries such as China's petrochemical, chemical, and steel industries were accelerated. The goal of resolving the upper limit of 150 million tonnes of excess steel production capacity in the "13th Five-Year Plan" was completed two years ahead of schedule, and more than 100 million tonnes of "illegal steel" production capacity was completely banned. China will continue to focus on key industries to implement effective control of greenhouse gas emissions, strengthen the management of energy consumption and carbon emission targets in key industries such as steel, building materials, chemicals, and non-ferrous metals, and promote the green transformation of industry. Based on this, various government departments in China have successively put forward the development plans for key industries during

Table 5 Policies for key areas under China's "Dual Carbon" goals

	Date	Published by	Document name
The action for green and low-carbon energy transition action	November 15, 2021	Five departments including the National Development and Reform Commission	Energy Efficiency Benchmarks and Baseline Levels in Key Areas of High-Energy-Consuming Industries (2021 Edition)
	January 30, 2022	National Development and Reform Commission, National Energy Administration	Opinions on Improving the System, Mechanism and Policy Measures for Green and Low-Carbon Energy Transformation
	March 22, 2022	National Development and Reform Commission, National Energy Administration	Notification on the "14th Five-Year Plan" Modern Energy System Plan
	March 23, 2022	National Development and Reform Commission, National Energy Administration	Medium- and Long-Term Plan for the Development of the Hydrogen Energy Industry (2021-2035)
	May 10, 2022	National Development and Reform Commission	Benchmark Levels and Baseline Levels in Key Areas of Clean and Efficient Utilization of Coal (2022 Edition)
	June 1, 2022	National Development and Reform Commission and nine other departments	"14th Five-Year Plan" for the Development of Renewable Energy
	September 20, 2022	National Energy Administration	Energy Carbon Peaking and Carbon Neutrality Standardization Enhancement Action Plan
The action for energy saving, carbon emission reduction and efficiency improvement action	June 1, 2021	State Administration of State Administration of Foreign Exchange, National Development and Reform Commission	Work Plan for Energy Conservation and Resource Conservation in Public Institutions During the 14th Five-Year Plan Period
	January 24, 2022	State Council	Comprehensive Work Plan for Energy Conservation and Emission Reduction during the 14th Five-Year Plan Period

(continued)

Table 5 (continued)

	Date	Published by	Document name
	February 11, 2022	Four departments including the National Development and Reform Commission	Implementation Guide for the Transformation and Upgrading of Energy Conservation and Carbon Reduction in Key Areas of High Energy-consuming Industries (2022 Edition)
	June 17, 2022	Seven departments including the Ministry of Ecology and Environment	Implementation Plan for Synergistic Pollution and Carbon Reduction
	October 18, 2022	Five departments including the National Development and Reform Commission	Several Opinions on Strict Energy Efficiency Constraints to Promote Energy Conservation and Carbon Reduction in Key Areas
	November 10, 2022	Five departments including the National Development and Reform Commission	Advanced Energy Efficiency Level, Energy Saving Level and Access Level for Key Energy-consuming Products and Equipment (2022 Edition)
The action for carbon peaking in the Industry sector	December 3, 2021	Ministry of Industry and Information Technology	"14th Five-Year Plan" for Industrial Green Development
	January 30, 2022	Nine departments including the Ministry of Industry and Information Technology	"14th Five-Year Plan" for the Development of the Pharmaceutical Industry
	February 7, 2022	Ministry of Industry and Information Technology, National Development and Reform Commission, Ministry of Ecology and Environment	Guiding Opinions on Promoting High-Quality Development of the Steel Industry
	April 7, 2022	Six departments including the Ministry of Industry and Information Technology	"Guiding Opinions on Promoting High-Quality Development of the Petrochemical Industry during the 14th Five-Year Plan Period"
	April 21, 2022	Ministry of Industry and Information Technology, National Development and Reform Commission	Guiding Opinions on the High-Quality Development of the Chemical Fiber Industry Guiding Opinions on the High-Quality Development of the Industrial Textiles Industry

(continued)

Table 5 (continued)

	Date	Published by	Document name
	June 17, 2022	Five departments including the Ministry of Industry and Information Technology	"Guiding Opinions on Promoting the High-Quality Development of Light Industry"
	June 21, 2022	Six departments including the Ministry of Industry and Information Technology	Industrial Water Efficiency Improvement Action Plan
	June 29, 2022	Six departments including the Ministry of Industry and Information Technology	Industrial Energy Efficiency Improvement Action Plan
	August 1, 2022	Ministry of Industry and Information Technology, National Development and Reform Commission, Ministry of Ecology and Environment	Communication on Issuing the Implementation Plan for Carbon Peaking in the Industrial Sector
	August 22, 2022	Seven departments including the Ministry of Industry and Information Technology	Green and Low-carbon Development Action Plan for the Information and Communication Industry (2022-2025)
	November 10, 2022	Three departments including the Ministry of Industry and Information Technology	Implementation Plan for Carbon Peaking in the Non-ferrous Metals Industry
The action for carbon peaking in urban and rural development	October 21, 2021	General Office of the CPC Central Committee, General Office of the State Council	Opinions on Promoting Green Development of Urban and Rural Construction
	January 19, 2022	Ministry of Housing and Urban-Rural Development	"Communication on the 14th Five-Year Plan for the Development of the Construction Industry"
The action for carbon peaking in urban and rural development	February 11, 2022	State Council	"14th Five-Year Plan for Promoting Agricultural and Rural Modernization"

(continued)

Table 5 (continued)

	Date	Published by	Document name
	March 1, 2022	Ministry of Housing and Urban-Rural Development	"14th Five-Year Plan for the Development of Science and Technology for Housing and Urban-Rural Construction"
	March 11, 2022	Ministry of Housing and Urban-Rural Development	"14th Five-Year Plan" Building Energy Conservation and Green Building Development Plan
	June 30, 2022	Ministry of Agriculture and Rural Affairs, National Development and Reform Commission	Implementation Plan for Emission Reduction and Carbon Sequestration in Agriculture and Rural Areas
	July 13, 2022	Ministry of Housing and Urban-Rural Development, National Development and Reform Commission	Implementation Plan for Carbon Peaking in Urban and Rural Construction
	November 2, 2022	Ministry of Industry and Information Technology and other four departments	Carbon Peaking Plan for the Building Materials Industry
The actions for green and low-carbon transport	January 18, 2022	State Council	"14th Five-Year Plan" for the Development of a Modern Comprehensive Transportation System
	January 21, 2022	Ministry of Transport	Communication on the 14th Five-Year Plan for the Development of Green Transportation
	June 24, 2022	Ministry of Transport, National Railway Administration, Civil Aviation Administration of China, State Post Bureau	"Opinions of the CPC Central Committee and the State Council on fully and accurately implementing the new development concept and doing a good job in carbon peaking and carbon neutrality"
The actions for carbon emissions reduction through circular economy	July 1, 2021	National Development and Reform Commission	"14th Five-Year Plan" for the Development of the Circular Economy
	February 10, 2022	Ministry of Industry and Information Technology and other eight departments	"Implementation Plan on Accelerating the Comprehensive Utilization of Industrial Resources"

(continued)

Table 5 (continued)

	Date	Published by	Document name
The actions for encouraging innovation on green and low-carbon science and technology	August 18, 2022	Eight departments including the Ministry of Industry and Information Technology	Implementation Plan for Accelerating the Comprehensive Utilization of Industrial Resources
	April 2, 2022	National Energy Administration, Ministry of Science and Technology	" 14th Five-Year Plan for Scientific and Technological Innovation in the Energy Sector"
The actions for consolidating and enhancing carbon sink capacity	December 31, 2021	State Administration for Market Regulation	National Standardization Administration " Guidelines for the Approval and Verification of Forestry Carbon Sink Projects" (GB/T 41198-2021)
	February 21, 2022	Ministry of Natural Resources	Methods for calculating the economic value of marine carbon sinks
The actions for green and low-carbon lifestyle nationwide	May 7, 2022	Ministry of Education	"Work Plan for Strengthening the Construction of Higher Education Talent Training System for Carbon Peaking and Carbon Neutrality"
The actions for reaching carbon emissions peak by different regions in a structured and orderly manner	See Sect. 4 of this chapter for details		

the "14th Five-Year Plan" period and incorporated the "carbon peaking and carbon neutrality" goals into them. The relevant policy plans for key industries under China's "carbon peaking and carbon neutrality" goals are shown in Table 6.

c. Support System for China's "Dual Carbon" Goals

The carbon peaking and carbon neutrality plans require support plans from multiple aspects such as technology innovation, fiscal, financial and pricing systems, carbon emission standard measurement systems, and the green and low-carbon actions of the whole nation, as shown in Table 7.

3.3 *Impacts of China's "Dual Carbon" Commitments*

China's goal of achieving carbon neutrality by 2060 is the most significant climate commitment made by any country so far in terms of reducing the expected global

Table 6 Policies for key industries under China's "Dual Carbon" goals

	Date	Published by	Document name
Medicine	January 30, 2022	Nine departments including the Ministry of Industry and Information Technology	"14th Five-Year Plan for the Development of the Pharmaceutical Industry"
Steel	February 7, 2022	Ministry of Industry and Information Technology, National Development and Reform Commission, Ministry of Ecology and Environment	Guiding Opinions on Promoting High-Quality Development of the Steel Industry
Hydrogen energy	March 23, 2022	National Development and Reform Commission, National Energy Administration	Medium- and Long-Term Development Plan for Hydrogen Energy Industry (2021–2035)
Petrochemical	April 7, 2022	Six departments including the Ministry of Industry and Information Technology	"Guiding Opinions on Promoting High-Quality Development of the Petrochemical Industry during the 14th Five-Year Plan Period"
Chemical fiber industry	April 21, 2022	Ministry of Industry and Information Technology, National Development and Reform Commission	Guiding Opinions on the High-Quality Development of the Chemical Fiber Industry
Textiles	April 21, 2022	Ministry of Industry and Information Technology, National Development and Reform Commission	Guiding Opinions on the High-Quality Development of the Industrial Textiles Industry
Coal	May 10, 2022	National Development and Reform Commission	"Benchmark Levels and Baseline Levels in Key Areas of Clean and Efficient Utilization of Coal (2022 Edition)"
Renewable energy	June 1, 2022	Nine departments including the National Development and Reform Commission	14th Five-Year Plan for the Development of Renewable Energy
Light industry	June 17, 2022	Five departments including the Ministry of Industry and Information Technology	"Guiding Opinions on Promoting the High-Quality Development of Light Industry"
Information and communication	August 22, 2022	Seven departments including the Ministry of Industry and Information Technology	Action Plan for Green and Low-Carbon Development of the Information and Communication Industry (2022–2025)

(continued)

Table 6 (continued)

	Date	Published by	Document name
Non-ferrous metals	November 10, 2022	Three departments including the Ministry of Industry and Information Technology	Implementation Plan for Carbon Peaking of the Non-ferrous Metals Industry
Building materials	November 8, 2022	Four departments including the Ministry of Industry and Information Technology	Implementation Plan for Carbon Peaking of the Building Materials Industry

warming. It plays a crucial role in promoting global climate governance. This commitment is in line with China's goal of high-quality development and with the sustainable development goals of both China and all of humanity. President Xi Jinping's series of important speeches and the decision-making and deployment of the CPC Central Committee have drawn a magnificent blueprint and pointed out the pathway and direction for promoting climate and environmental governance and sustainable development. They demonstrate China's strategic determination to adhere to green and low-carbon development, as well as its responsibility as a major country in actively responding to climate change and promoting the building of a community with a shared future for mankind, which has won high praise and wide acclaim from the international community. China's carbon peaking and carbon neutrality strategies are not only major requirements for global climate governance, protecting the Earth's homeland, and building a community with a shared future for mankind, but also internal needs for China's high-quality development, ecological civilization construction, and comprehensive ecological and environmental governance.

As the largest developing country, China has taken the initiative to assume international responsibilities for addressing climate change, setting a good example for other countries to strengthen their climate actions and promoting the establishment of a global climate governance system to enter a new era. Currently, the global economic landscape is being reshaped. The COVID-19 pandemic has triggered a resurgence of unilateralism, drawing more attention to the prospects of global climate governance. As the country with the largest carbon emissions in the world, China's proactive proposal of the carbon neutrality goal despite the slowdown in global economic growth has effectively countered the impacts brought about by unilateralism and de-globalization. It is of great significance for promoting the achievement of the goals of the Paris Agreement (Ruan Yunzhi, 2020). From a participant in global climate governance to a contributor and then to a leader, the change in China's role in the field of global climate governance is not only an internal need to safeguard national interests and promote domestic sustainable development, but also a manifestation of the responsibility of a major developing country and a historical mission to promote the building of a community with a shared future for mankind.

Achieving carbon peaking and carbon neutrality is a major strategic decision made by the CPC Central Committee and the State Council, taking into account both the

Table 7 Policy support system for China's "Dual Carbon" goals

	Date	Published by	Document name
The actions for green and low-carbon technological innovation	April 2, 2022	National Energy Administration, Ministry of Science and Technology	"14th Five-Year Plan for Scientific and Technological Innovation in the Energy Sector"
	August 18, 2022	Nine departments including the Ministry of Science and Technology	"Implementation Plan for Science and Technology Support for Carbon Peaking and Carbon Neutrality (2022–2030)"
The actions for green and low-carbon lifestyle nation wide	May 7, 2022	Ministry of Education	"Work Plan for Strengthening the Construction of Higher Education Talent Training System for Carbon Peaking and Carbon Neutrality"
Carbon Emissions Statistics and Accounting System	March 15, 2022	General Office of the Ministry of Ecology and Environment	Notification on the key tasks related to the management of corporate greenhouse gas emissions reporting in 2022
	October 18, 2022	Nine departments including the State Administration for Market Regulation	"Communication on Issuing the Implementation Plan for Establishing and Improving the Standard and Measurement System for Carbon Peaking and Carbon Neutrality"
	December 19, 2022	CPC Central Committee, State Council	Guiding Opinions on Building a Data Infrastructure System to Better Play the Role of Data Elements
Fiscal and financial system for "carbon peaking and carbon neutrality"	November 27, 2021	State-owned Assets Supervision and Administration Commission	"Guiding Opinions on Promoting the High-Quality Development of Central Enterprises and Doing a Good Job in Carbon Peaking and Carbon Neutrality"
	March 1, 2022	Shanghai Stock Exchange	Action Plan for Carbon Peaking and Carbon Neutrality During the 14th Five-Year Plan Period of the Shanghai Stock Exchange
	May 13, 2022	China Banking and Insurance Regulatory Commission	Guidelines for Green Finance in the Banking and Insurance Industries

(continued)

Table 7 (continued)

	Date	Published by	Document name
	May 31, 2022	Ministry of Finance	Opinions on Fiscal Support for Doing a Good Job in Carbon Peaking and Carbon Neutrality
	May 31, 2022	State Administration of Taxation	Guidelines on Preferential Tax and Fee Policies to Support Green Development

international and domestic situations. It is of great significance for China's ecological civilization construction, leading global climate change governance, and achieving the "Two Centenary Goals". The "30·60 Goals" demonstrate China's firm determination to follow the pathway of green and low-carbon development, contribute China's wisdom and solutions to countries around the world in joining hands to address global challenges and jointly protect the Earth's homeland, reflect China's responsibility as a major country in taking the initiative to assume international responsibilities for addressing climate change and promoting the building of a community with a shared future for mankind, fully showcase to the international community China's determination and confidence as a responsible major country in pursuing green and low-carbon development and building a clean and beautiful world, and have won high recognition from the international community. The British Broadcasting Corporation (BBC) called it "a very inspiring step". The New York Times of the United States said that if China realizes this commitment, it will make a significant contribution to slowing down global warming. In addition, overseas media in Russia, Greece, Switzerland and other countries also gave high praise to China's carbon neutrality goal.[4]

4 The "Dual Carbon" Goals, Strategies and Actions of China's Provinces and Autonomous Regions

4.1 Overview of Carbon Neutrality Practices in Provinces and Regions in China

a. Regions Actively Respond to National Carbon Neutrality Goals

Since 2021, in order to promote the effort for carbon peaking and carbon neutrality, many provinces have issued their own "Implementation Opinions on Fully, Accurately and Comprehensively Implementing the New Development Concept and

[4] China Economic Net. China advocates pointing the way forward for the international community—overseas media highly praised President Xi Jinping's important speech at the series of high-level meetings on the 75th anniversary of the founding of the United Nations [EB/OL] (2020-09-26) [2023-11-02].

Doing a Good Job in Carbon Peak and Carbon Neutrality". Most parts of China have actively responded to the national carbon neutrality policy in the "14th Five-Year Plan" and the Government Work Report. So far, 19 provinces (autonomous regions, municipalities directly under the central government) have officially issued carbon peaking implementation plans, including Beijing, Shanghai, Tianjin, Jiangsu, Hunan and others. The "Dual Carbon" implementation plans of some regions are shown in Table 8.

b. Regions Set Carbon Peaking Targets by Industry

The deployment of carbon peaking policies mainly focuses on areas such as energy, construction, transportation, ecology, and marketing.

In the energy sector, most regions across the country have proposed to adhere to and improve the system of dual control mechanism (energy intensity & total consumption), and promote the adjustment of the energy structure, including controlling the total amount of coal and increasing the proportion of non-fossil energy in primary energy. Among them, Shanghai, Zhejiang, Hainan, Anhui, Xinjiang, Ningxia, Henan, etc. have set specific quantitative goals for the energy industry. For example, Shanghai plans to control the total coal consumption at around 43 million tonnes by 2025, reduce the proportion of coal consumption in primary energy consumption to around 30%, increase the proportion of natural gas in primary energy consumption to around 17%, and increase the proportion of local renewable energy in the total electricity consumption of the whole society to around 8%.

In the construction sector, building energy conservation is the direction that most regions mention in their policy planning, but there are significant differences in the detail and power of the goals. Among them, Jiangsu has formulated the most detailed plan for the construction industry. By 2025, the total scale of green buildings in the province will remain the largest in the country. On the basis of a 30% increase in energy conservation for new buildings in 2020, it will be further increased by 30%. The carbon emission intensity of buildings will strive to be the lowest in the country. By 2030, the carbon peaking task in the field of housing and urban–rural construction will be achieved. Qinghai has also set specific goals for the construction industry, that is, to accelerate the green transformation of existing buildings, promote green building materials and prefabricated buildings, and increase the proportion of green buildings in newly constructed buildings to 70% by 2025. Other regions, such as Chongqing, Gansu, Fujian, Shanghai, etc., have put forward qualitative policy measures such as developing green buildings, accelerating the development of prefabricated buildings, and improving the energy-saving and low-carbon level of buildings.

In the transportation sector, the development of new energy vehicles is an important focus for energy conservation and emission reduction in the transportation field. Hainan, Guizhou, Gansu, Shanghai, Fujian and other regions have put forward specific requirements for the number of new energy vehicles, the scale of newly built charging infrastructure, and the proportion of new energy vehicles in the public domain. For example, Hainan will promote 25,000 clean energy vehicles, and 100% of newly added and replaced vehicles in the public domain will be clean energy

Table 8 Implementation plans for "Dual Carbon" in some provinces (autonomous regions, municipalities directly under the central government) of China

Province (autonomous regions, municipalities directly under the central government)	Date	Published by	Document name
Beijing	October 11, 2022	The People's Government of Beijing Municipality	"Implementation Plan for Carbon Peaking in Beijing"
	June 7, 2023	Office of the Leading Group for Carbon Peaking and Carbon Neutrality Work in Beijing	"Action Plan for Renewable Energy Substitution in Beijing (2023–2025)"
Fujian Province	August 22, 2022	The Communist Party of China Fujian Provincial Committee, The People's Government of Fujian Province	"Implementation Opinions on Fully, Accurately and Comprehensively Implementing the New Development Concept and Doing a Good Job in Carbon Peaking and Carbon Neutrality"
Guangdong Province	July 25, 2022	The Communist Party of China Guangdong Provincial Committee, The People's Government of Guangdong Province	"Implementation Opinions on Fully, Accurately and Comprehensively Implementing the New Development Concept and Promoting the Work of Carbon Peaking and Carbon Neutrality"
	February 7, 2023	The People's Government of Guangdong Province	"Implementation Plan for Carbon Peaking in Guangdong Province"
Guangxi Zhuang Autonomous Region	May 13, 2022	The Communist Party of China Guangxi Zhuang Autonomous Region Committee, The People's Government of Guangxi Zhuang Autonomous Region	"Implementation Opinions on Fully, Accurately and Comprehensively Implementing the New Development Concept and Doing a Good Job in Carbon Peaking and Carbon Neutrality"
Guizhou Province	November 4, 2022	The People's Government of Guizhou Province	"Implementation Plan for Carbon Peaking in Guizhou Province"
Hainan Province	August 22, 2022	The People's Government of Hainan Province	"Implementation Plan for Carbon Peaking in Hainan Province"

(continued)

Table 8 (continued)

Province (autonomous regions, municipalities directly under the central government)	Date	Published by	Document name
Hebei Province	January 5, 2022	The Communist Party of China Hebei Provincial Committee, The People's Government of Hebei Province	"Implementation Opinions on Fully, Accurately and Comprehensively Implementing the New Development Concept and Doing a Good Job in Carbon Peaking and Carbon Neutrality"
	February 28, 2023	Hebei Provincial Department of Science and Technology, Hebei Provincial Development and Reform Commission, Hebei Provincial Department of Industry and Information Technology, Hebei Provincial Department of Ecology and Environment, etc	"Implementation Plan for Scientific and Technological Support for Carbon Peaking and Carbon Neutrality in Hebei Province (2023–2030)"
Henan Province	February 6, 2023	The Communist Party of China Henan Provincial Committee, The People's Government of Henan Province	"Implementation Plan for Carbon Peaking in Henan Province"
	February 23, 2022	The People's Government of Henan Province	"Plan for the Modern Energy System and Carbon Peaking and Carbon Neutrality in Henan Province During the 14th Five-Year Plan Period"
Hunan Province	March 22, 2022	The Communist Party of China Hunan Provincial Committee, Hunan Provincial People's Government	"Implementation Opinions on Fully, Accurately and Comprehensively Implementing the New Development Concept and Doing a Good Job in Carbon Peaking and Carbon Neutrality"

(continued)

Table 8 (continued)

Province (autonomous regions, municipalities directly under the central government)	Date	Published by	Document name
	October 8, 2022	The Communist Party of China Hunan Provincial Committee, The People's Government of Hunan Province	"Implementation Opinions on Fully, Accurately and Comprehensively Implementing the New Development Concept and Doing a Good Job in Carbon Peaking and Carbon Neutrality"
Jilin Province	November 30, 2021	The Communist Party of China Jilin Provincial Committee, The People's Government of Jilin Province	"Implementation Opinions on Fully, Accurately and Comprehensively Implementing the New Development Concept and Doing a Good Job in Carbon Peaking and Carbon Neutrality"
	August 1, 2022	The People's Government of Jilin Province	"Implementation Plan for Carbon Peaking in Jilin Province"
Jiangsu Province	See Sect. 4, Part 2 of this chapter for details		
Jiangxi Province	April 6, 2022	The Communist Party of China Jiangxi Provincial Committee, The People's Government of Jiangxi Province	"Implementation Opinions on Fully, Accurately and Comprehensively Implementing the New Development Concept and Doing a Good Job in Carbon Peaking and Carbon Neutrality"
	July 18, 2022	The People's Government of Jiangxi Province	"Implementation Plan for Carbon Peaking in Jiangxi Province"
Inner Mongolia Autonomous Region	June 28, 2022	The Party Committee of Inner Mongolia Autonomous Region, The People's Government of Inner Mongolia Autonomous Region	"Implementation Opinions on Fully, Accurately and Comprehensively Implementing the New Development Concept and Doing a Good Job in Carbon Peaking and Carbon Neutrality"

(continued)

Table 8 (continued)

Province (autonomous regions, municipalities directly under the central government)	Date	Published by	Document name
	November 17, 2022	The People's Government of Inner Mongolia Autonomous Region	"Implementation Plan for Carbon Peaking in Inner Mongolia Autonomous Region"
Shanghai	July 28, 2022	The People's Government of Shanghai Municipality	"Implementation Plan for Carbon Peaking in Shanghai Municipality"
Sichuan Province	March 31, 2022	The Communist Party of China Sichuan Provincial Committee, The People's Government of Sichuan Province	"Implementation Opinions on Fully, Accurately and Comprehensively Implementing the New Development Concept and Doing a Good Job in Carbon Peaking and Carbon Neutrality"
	December 31, 2022	The People's Government of Sichuan Province	"Implementation Plan for Carbon Peaking in Sichuan Province"
Tianjin	August 25, 2022	The People's Government of Tianjin Municipality	"Implementation Plan for Carbon Peaking in Tianjin Municipality"
Zhejiang Province	February 17, 2022	The Communist Party of China Zhejiang Provincial Committee, The People's Government of Zhejiang Province	"Implementation Opinions on Fully, Accurately and Comprehensively Implementing the New Development Concept and Doing a Good Job in Carbon Peaking and Carbon Neutrality"
	February 21, 2023	The Department of Economy and Information Technology of Zhejiang Province, The Development and Reform Commission of Zhejiang Province, The Department of Ecology and Environment of Zhejiang Province	"Implementation Plan for Carbon Peaking in the Industrial Field of Zhejiang Province"

(continued)

Table 8 (continued)

Province (autonomous regions, municipalities directly under the central government)	Date	Published by	Document name
Chongqing	January 26, 2021	Chongqing Municipal Ecology and Environment Bureau	"Technical Guidelines for Environmental Impact Assessment of Planning in Chongqing Municipality - Carbon Emission Assessment"
	February 15, 2022	General Office of the People's Government of Chongqing Municipality, General Office of the People's Government of Sichuan Province	"Joint Action Plan for Carbon Peaking and Carbon Neutrality in the Chengdu-Chongqing Economic Circle"
	July 9, 2022	The People's Government of Chongqing Municipality	"Action Plan for Deeply Promoting the High-quality and Green Development of the Manufacturing Industry Guided by the Goals of Carbon Peaking and Carbon Neutrality (2022 - 2025)"
	July 29, 2022	The Communist Party of China Chongqing Municipal Committee, The People's Government of Chongqing Municipality	"Implementation Opinions on Fully, Accurately and Comprehensively Implementing the New Development Concept and Promoting the Work of Carbon Peaking and Carbon Neutrality"
	January 28, 2023	The Chongqing Municipal Commission of Economy and Information Technology, The Chongqing Municipal Development and Reform Commission, The Chongqing Municipal Department of Ecology and Environment	"Implementation Plan for Carbon Peaking in the Industrial Field of Chongqing Municipality"
Chongqing	August 27, 2023	Chongqing Municipal Ecology and Environment Bureau	"Implementation Plan for Synergistic Pollution and Carbon Reduction in Chongqing Municipality"

(continued)

Table 8 (continued)

Province (autonomous regions, municipalities directly under the central government)	Date	Published by	Document name
Qinghai Province	December 19, 2022	The People's Government of Qinghai Province	"Implementation Plan for Carbon Peaking in Qinghai Province"
Yunnan Province	May 21, 2022	The Communist Party of China Yunnan Provincial Committee, The People's Government of Yunnan Province (Issued)	"16 Key Measures for Yunnan Province to Strive to Become a Pioneer in the Construction of Ecological Civilization"
	August 11, 2022	The People's Government of Yunnan Province	"Implementation Plan for Carbon Peaking in Yunnan Province"

vehicles. The overall vehicle-to-pile ratio of electric vehicles and charging infrastructure will be below 2.5:1. Guizhou aims to build more than 20,000 new charging infrastructure by 2025 and increase the charging guarantee capacity of new energy electric vehicles to 3.5 kilowatts per vehicle.

In terms of coordinated pollution control, the coordinated control of greenhouse gases, air pollution control, water pollution control, and domestic waste treatment are also the focuses of many regions. Beijing, Shanghai, Zhejiang, Ningxia, Guizhou and other places have set up relatively high standards. For example, Zhejiang plans to implement coordinated governance measures for greenhouse gases and pollutants, aiming to achieve that the proportion of days with excellent air quality in cities divided into districts reaches above 93%, the proportion of surface water with water quality of Class III and above reaches above 95%, and all cities divided into districts and 60% of counties (cities) will build "waste-free cities".

In terms of ecological construction, most regions actively respond to the construction of national ecological civilization, and some regions further identified specific requirements for afforestation and water quality. For example, Gansu actively promotes the industrial ecologicalization and ecological industrialization. Shaanxi plans to create a national ecological civilization pilot area. Zhejiang and Guizhou plan to increase their forest coverage rates to 61.5% and 60% respectively. Anhui and Liaoning aim to have afforestation areas of 1.4 million mu and 2.02 million mu respectively.

In terms of the construction of the marketing mechanism, with the acceleration of the construction process of the national carbon market, regions such as Shanxi, Henan, Heilongjiang, and Jilin have successively carried out basic capacity building such as carbon statistics, verification, and supervision, as well as the construction of regional carbon markets. Besides the carbon market, the carbon sink market, carbon certification, and climate investment and financing are also hotspots of concern in a lot of regions. Shanghai, Sichuan, Chongqing and other places have made relatively detailed plans for the above fields.

c. The Characteristic Positioning of Carbon Peaking of Different Regions

China has a vast territory, and there are significant differences in resource endowments and development levels among various regions, so there are also distinct regional characteristics in the work of carbon peaking and carbon neutrality.

(i) Energy Bases

Energy-rich regions plan and build energy bases with their own characteristics according to their own advantages. For example, Qinghai will be built into a demonstration province of clean energy and a pilot demonstration area for the development of national energy storage. Inner Mongolia plans to build an important national energy and strategic resource base. Xinjiang will be built into a national important clean energy base of "Three Bases and One Channel". Gansu will build an important modern comprehensive energy production base, reserve base, export base and strategic channel for the country. Shanxi plans to basically build a green energy supply system. Anhui will promote the construction of a green energy storage base.

(ii) Zero-Carbon Demonstration Areas

Regions with relatively rapid industrial development such as Zhejiang, Hubei, Sichuan, and Jilin plan to build green industrial bases and construct demonstration projects for "zero-carbon" emission areas.

(iii) Ecological Civilization Pilot Areas

Regions such as Fujian, Chongqing, and Shaanxi plan to create national ecological civilization pilot areas and further carry out pilot projects for low-carbon cities, low-carbon parks, and low-carbon communities.

(iv) Other Characteristic Areas

Shanghai will give play to the advantages of an international financial center and build an international carbon financial center. Heilongjiang will explore the development of carbon sink economy, strive to build a carbon sink trading center, and build a major province of carbon sink economy in the country.

4.2 The Practice of "Dual Carbon" in China's Provinces and Autonomous Regions: Using Jiangsu Province as an Example

The carbon neutrality practices within the provinces in China also follow the "1 + N" approach. This section will take Jiangsu Province as an example to specifically demonstrate how provincial-level regions in China construct a provincial-level policy system for carbon peaking and carbon neutrality under the national "1 + N" system, and introduce the policy documents related to key areas, key industries and the support and guarantee system in Jiangsu Province.

a. The "1 + N" Policy System for Dual Carbon in Jiangsu Province

The national "1 + N" policy system for carbon peaking and carbon neutrality includes the "carbon peaking and carbon neutrality" goals of different provinces, autonomous regions and municipalities. Similarly, within each province, autonomous region and municipality, there is also a "carbon peaking and carbon neutrality" policy system with a "1 + N" structure. On the one hand, Jiangsu Province has issued the provincial-level "1" policy document, that is, the overall guiding document. In January 2022, the Communist Party of China Jiangsu Provincial Committee and the People's Government of Jiangsu Province released the "Implementation Opinions on Promoting High-quality Development and Doing a Good Job in Carbon Peaking and Carbon Neutrality", proposing that the main goals of the province are to initially form an economic system of green, low-carbon and circular development and effectively control carbon emissions by 2025, to achieve remarkable results in the green and low-carbon transformation of the economy and society and improve the coordinated management system for pollution reduction and carbon reduction by 2030, and to fully establish an economic system of green, low-carbon and circular development and a clean, low-carbon, safe and efficient energy system by 2060, with the energy utilization efficiency reaching the advanced level internationally and achieving the carbon neutrality goal as scheduled. In October 2022, the People's Government of Jiangsu Province released the "Implementation Plan for Carbon Peaking in Jiangsu Province", proposing that the carbon dioxide emissions will peak before 2030, providing a strong support for achieving carbon neutrality. In addition, on November 25, 2022, the Standing Committee of the Jiangsu Provincial People's Congress passed the "Decision on Promoting Carbon Peaking and Carbon Neutrality", proposing to clarify the overall requirements for promoting carbon peaking and carbon neutrality and construct the working mechanism for promoting carbon peaking and carbon neutrality. All of the above governing documents have led the development of the "carbon peaking and carbon neutrality" goals in Jiangsu Province.

b. Key Areas and Key Industries for the "Dual Carbon" Goals in Jiangsu Province

In terms of key areas, the "Decision on Promoting Carbon Peaking and Carbon Neutrality" released by the Standing Committee of the Jiangsu Provincial People's Congress proposes to accelerate the green and low-carbon transformation in key areas, specifically including seven major areas: accelerating the low-carbon transformation in the energy field, accelerating the promotion of the green development of industries, effectively improving the level of energy conservation and efficiency improvement, comprehensively improving the quality of green and low-carbon development in urban and rural construction, continuously promoting the construction of a green and low-carbon transportation system, consolidating and enhancing the carbon sink capacity of the ecosystem, and actively promoting the coordinated and efficient pollution reduction and carbon reduction. In addition, the Communist Party of Jiangsu Provincial Committee of China and the People's Government of Jiangsu Province released the "Implementation Opinions on Promoting High-quality Development and Doing a Good Job in Carbon Peaking and Carbon Neutrality", which put forward eight specific requirements, including building a green and low-carbon transformation, a low-carbon and efficient industrial structure, a low-carbon and safe energy utilization system, a green and low-carbon transportation system, a low-carbon urban and rural construction and development system, a low-carbon innovative technology application system, an ecological carbon sink consolidation and enhancement system, and a supporting system for green and low-carbon transformation. In terms of key industries, the "Implementation Plan for Carbon Peaking in Jiangsu Province" published by the People's Government of Jiangsu Province points out that it will promote the carbon peaking actions of key industrial industries, including steel, petrochemical and chemical, building materials and other industries. By comparing this part with the key areas and key industries for China's "carbon peaking and carbon neutrality" goals in Section III of this chapter, it can be understood that Jiangsu Province has a high degree of overlap with the country.

c. Support and Guarantee for the "Dual Carbon" Goals in Jiangsu Province

The "Decision on Promoting Carbon Peaking and Carbon Neutrality" published by the Standing Committee of the Jiangsu Provincial People's Congress proposes to strengthen the establishment of the support and guarantee system for the "carbon peaking and carbon neutrality" goals, including six aspects such as effectively strengthening the innovation of green and low-carbon science and technology, continuously improving the policy support system, establishing and improving the market-oriented mechanism, accelerating the improvement of the statistical accounting system for carbon emissions and carbon sinks, deeply carrying out publicity and education on carbon peaking and carbon neutrality, and accelerating the formation of a scientific and efficient assessment and evaluation system. Similarly, by comparing it with the support and guarantee systems of national policies, it can be observed that they have similarities.

The policies and plans related to the "Dual Carbon" goals in Jiangsu Province are shown in Table 9.

4.3 The Relationship Between National and Regional Carbon Neutrality Goals

The carbon neutrality goals at the national and regional levels are in a relationship where the national goal plays a leading role and the regions are the main implementers. It is also a relationship of mutual complementarity and mutual promotion. Their common purpose is to reduce the country's carbon emissions and respond to global climate change. China's "30·60 Goals" (referring to achieving carbon peaking by 2030 and carbon neutrality by 2060) involve a tight schedule and heavy tasks. The

Table 9 Policies and plans related to the "Dual Carbon" goals in Jiangsu Province

Date	Published by	Document Name
Policies and Plans of the "1" Category in Jiangsu Province		
January 15, 2022	The Communist Party of Jiangsu Provincial Committee of China, The People's Government of Jiangsu Province	"Implementation Opinions on Promoting High-quality Development and Doing a Good Job in Carbon Peaking and Carbon Neutrality"
October 2, 2022	The People's Government of Jiangsu Province	"Implementation Plan for Carbon Peaking in Jiangsu Province"
Policies and Plans of the "*N*" Category in Jiangsu Province		
April 15, 2021	The Department of Housing and Urban–Rural Construction of Jiangsu Province	"Guiding Opinions on Promoting Green Urban and Rural Construction Under the Goal of Carbon Peaking"
January 12, 2023	The Department of Industry and Information Technology of Jiangsu Province, The Development and Reform Commission of Jiangsu Province, The Department of Ecology and Environment of Jiangsu Province	"Implementation Plan for Carbon Peaking in the Industrial Field of Jiangsu Province and Carbon Peaking in Key Industries"
January 13, 2023	The Department of Housing and Urban–Rural Construction of Jiangsu Province, The Development and Reform Commission of Jiangsu Province	"Implementation Plan for Carbon Peaking in the Field of Urban and Rural Construction in Jiangsu Province"
February 10, 2023	The Science and Technology Department of the Department of Transportation of Jiangsu Province	"Implementation Plan for Green and Low-carbon Development in the Field of Transportation in Jiangsu Province"
April 4, 2023	The Department of Finance of Jiangsu Province	"Implementation Strategies of Jiangsu Province's Fiscal Support for Doing a Good Job in Carbon Peaking and Carbon Neutrality"

realization of the "30·60 Goals" requires not only top-level design at the national level but also the full cooperation of all provinces, autonomous regions, and municipalities.

The national setting of carbon neutrality goals is to guide the society to transform towards low-carbon development. This not only reflects the macro-strategy at the national level but also provides clear development directions for organizations and projects. To achieve this goal, the country has formulated action plans and programs for achieving carbon peaking before 2030 and carbon neutrality before 2060. This involves overall design at the national level, including formulating strategies for coordinated regional development and differentiated low-carbon transformation strategies.

Regions, as the specific implementers of the national carbon neutrality operations, need to break down the national carbon neutrality goals into regional carbon neutrality goals and be responsible for their implementation. The carbon neutrality effects of each region need to be consistent with the national carbon neutrality goals, which is similar to the relationship between the global temperature rise target and the nationally determined contributions targets in the Paris Agreement. To ensure the achievement of the goals, it is necessary to regularly evaluate and update the carbon neutrality goals of each region. In addition, each region should also formulate scientific, comprehensive, industry-wide, and quantified roadmaps for carbon peaking and carbon neutrality as soon as possible to ensure that regions take effective and orderly actions in the process of achieving carbon peaking and carbon neutrality and ultimately reach the overall national goals.

From the perspective of the timeline, before the specific carbon neutrality plans at the national and industry levels are optimized, each region can take the lead in formulating carbon neutrality plans that conform to its own development directions to assist in the formulation of the national carbon neutrality goals. In particular, low-carbon pilot areas and economically developed regions should play an exemplary role. Regions can also first do a good job in infrastructure construction such as carbon inventory, and then formulate detailed and accurate regional carbon neutrality action plans after the national carbon neutrality plan is introduced.

Chapter 2
Carbon Neutrality is Not Equal to Carbon Offsetting—Critical Challenges in Pathways Toward Carbon Neutrality

Carbon neutrality is not equal to carbon offsetting. The core idea of carbon neutrality is to decouple economic development from carbon emissions. The impact of economic development on climate change can be minimized through low-carbon energy, advanced production processes and negative emission technologies. The concept of carbon neutrality does go beyond simple carbon offsetting. It emphasizes the reduction or elimination of carbon emissions in the process of economic development, so that there is no longer a positive correlation between economic growth and carbon emissions. This decoupling is achieved through the combined effects of multiple methods.

The core of carbon emission decoupling is that economic growth no longer relies on high-carbon energy, but is achieved through the promotion of low-carbon energy and the construction of a clean energy system. While the economy is growing, if the growth rate of carbon dioxide emissions is negative or less than the growth rate of the economy, this can be regarded as decoupling. Its essence is to measure whether economic growth comes at the cost of resource depletion and environmental degradation. Carbon emission decoupling is an idealized process in which the connection between economic growth and greenhouse gas emissions continues to weaken or even disappear, that is, gradually reduce the consumption of high-carbon energy sources while maintaining the level economic growth. This transformation means that while the economy continues to grow, the growth rate of carbon dioxide emissions will slow down or even decrease, achieving the decoupling between economic growth and greenhouse gas emissions.

The utilization of low-carbon energy is an important way to decouple carbon emissions, enabling "relative decoupling" of carbon emissions. Developing clean energy resources, including wind energy, solar energy, nuclear energy, geothermal energy and biomass energy, to replace fossil energy such as coal and oil, helps reduce carbon dioxide emissions. In the field of energy, the goal is to build a clean, low-carbon, safe and efficient energy system, which involves controlling the total amount of fossil energy, improving utilization efficiency, implementing renewable

C. Chen et al., *Sustainable Carbon Neutrality*,
https://doi.org/10.1007/978-981-95-1691-9_2

energy substitution actions, deepening power system reform, and building a new power system with new energy as the main body. The utilization of low-carbon energy is the key to achieving this goal. By vigorously developing clean energy such as wind energy, solar energy, nuclear energy, geothermal energy and biomass energy, we can effectively reduce dependence on fossil fuels such as coal and oil, thereby reducing carbon emissions in the process of energy production and consumption. These clean energy sources not only help reduce greenhouse gas emissions, but also improve energy security and sustainability. In order to achieve the widespread utilization of low-carbon energy, it is necessary to build a clean, low-carbon, safe and efficient energy system. This includes controlling the total amount of fossil energy, improving energy utilization efficiency, implementing renewable energy substitution actions, and power system reforming to a deeper level. Through these measures, we can gradually build a new power system with new energy as the main body, and improve the stability and reliability of energy supply.

The goal of achieving decoupling of carbon emissions is to ensure that economic growth is no longer achieved at the expense of the environment and resources, but rather through innovation and transformation that can promote the coordinated and sustainable development of the economy, environment and society. This process involves not only technological innovation, such as improving energy efficiency and adopting clean energy technologies, but also optimizing the economic structure, such as developing low-carbon industries and promoting green finance. Improving social awareness, such as consumer demand for sustainable products, and policy support, such as carbon pricing mechanisms and environmental regulations, are indispensable factors in achieving decoupling.

Carbon emission decoupling also emphasizes energy-saving management on the energy demand side. This requires us to use advanced technology to improve production processes, improve the efficiency of energy resource utilization, and pay special attention to improving the energy efficiency of energy-consuming equipment. In the industrial sector, it is crucial to implement green manufacturing, which includes adopting environmentally friendly materials, optimizing production processes and reducing waste generation. The construction sector needs to raise energy-saving standards and reduce energy consumption through green building design, high-performance insulation materials and intelligent building management systems. The transportation sector should also develop low-carbon transportation modes, such as promoting electric vehicles, raising fuel efficiency standards and encouraging the use of public transportation. In addition, energy demand can be further reduced by raising public awareness of energy conservation and promoting energy-saving lifestyles. Policymakers can facilitate the increase in energy efficiency by formulating and enforcing energy-saving standards, providing tax incentives and subsidies. Enterprises and organizations should also invest in research and development to find innovative solutions to reduce energy consumption and carbon emissions.

Negative emission technologies play a vital role in achieving carbon neutrality because they remove carbon dioxide from the atmosphere, rather than just reducing emissions. These technologies include afforestation and reforestation, carbon capture, utilization and storage (CCUS) technologies, and methods such as enhancing marine carbon sinks. Afforestation and reforestation absorb carbon dioxide from the atmosphere through the photosynthesis of plants and store it in plant biomass and soil. This not only increases carbon sinks, but also benefits biodiversity protection and sustainable land management. Carbon capture, utilization and storage technologies capture carbon dioxide directly from industrial emission sources or the atmosphere and store it safely underground, thereby reducing the amount of greenhouse gases in the atmosphere. This technology is essential for achieving carbon neutrality in industries in which it is difficult to reduce emissions. Ocean carbon sinks help absorb and store carbon dioxide through marine activities, such as the carbon cycling process of marine organisms. This can be enhanced by methods such as marine fertilization and seaweed farming. To enhance the carbon sink capacity of ecosystems, it is necessary to strengthen land space planning and land use control to ensure that natural ecosystems such as forests, grasslands, wetlands, oceans, soils and permafrost are effectively protected and rationally used. This includes developing and implementing protection measures, restoring degraded ecosystems, and promoting sustainable land management practices. These negative emission technologies and ecological protection measures can further increase the carbon sinks in the ecosystem, contributing to the achievement of global carbon neutrality goals.

The goal of carbon emission decoupling is to achieve the decoupling of economic growth and environmental impact through these comprehensive measures, thus promoting a more sustainable and environmentally friendly social and economic development model. In short, carbon emission decoupling is a multifaceted system project that requires the joint efforts of the governments, enterprises and all sectors of society. Therefore, carbon neutrality is not just the concept of carbon offsetting. The most important thing is voluntary emission reduction. Carbon neutrality promotes technological transformation, upgrading and high-quality development of industry and energy.

1 Determination of Scope

1.1 Boundary

Determining the carbon neutrality boundary is the basis for implementing the carbon neutrality commitments, which requires accurate identification and classification of carbon emission sources and carbon sinks to ensure clear boundaries and avoid double counting. For countries, provinces, cities and other regions, boundaries are usually determined based on administrative divisions, while at the organizational level, they may be specific to streets, house numbers and buildings. The boundaries of product carbon neutrality are determined based on product units, covering a series of products with the same design, functional structure, materials and processes. For energy-consuming products, the entire life cycle is considered, while for non-energy-consuming products, the cradle-to-gate approach is adopted. At the same time, it should be clearly stated which venues and activities are not included in the carbon neutrality boundary. The determination of boundaries requires dynamic management to adapt to changes in policies, technologies and other factors, ensuring that the realization of carbon neutrality goals is both scientific and fair.

1.2 Types of Greenhouse Gases

Determine whether the types of greenhouse gases (GHGs) for carbon neutrality should include greenhouse gases other than carbon dioxide. The types of greenhouse gases selected for regional carbon neutrality projects should be consistent with the categories outlined in the national carbon neutrality goals to ensure the successful achievement of national carbon neutrality goals. At the organizational, product and project levels, the types of greenhouse gases can be determined based on the carbon neutrality mechanism of their corresponding services. If the types of greenhouse gases controlled are less than the range of greenhouse gases determined by the carbon neutrality mechanism, this should be made clear and justified in self-declarations, letters of commitment, reports or certificates.

According to the IPCC Fifth Assessment Report, the types of human-induced greenhouse gases that need to be controlled are carbon dioxide, methane, nitrous oxide, hydrofluorocarbons, perfluorocarbons, sulfur hexafluoride and nitrogen trifluoride. Carbon dioxide accounts for the largest proportion of all man-made greenhouse gases, more than 70%,[1] and is also the main "culprit" causing the greenhouse effect. Methane, nitrous oxide and other greenhouse gases account for approximately 15%, 6%, and 2%, respectively.

The sources of carbon dioxide emissions are divided into three categories: carbon dioxide released from energy (combustion), carbon dioxide released from other industrial production (non-combustion, such as carbon dioxide released from the calcium carbonate in lime production, etc.), and carbon dioxide released from agriculture and other activities. The largest proportion of carbon dioxide emissions comes from the energy sector.

Methane is the second largest source of anthropogenic emissions after carbon dioxide, and its greenhouse gas effect is about 25 times greater than that of carbon dioxide. According to the observations of[2] the IPCC, the global average concentration of methane has increased by about 80% over the past 100 years. The main sources of anthropogenic methane emissions are the coal industry, agriculture (rice planting and breeding industry), oil and gas industry, waste treatment, etc.

Nitrous oxide mainly comes from agriculture, accounting for about 70% of the total. The growth of human demand for food and animal feed will further increase the global nitrous oxide emissions. The production of nitric acid and adipic acid in

[1] World Resources Institute research data, 2014 statistical values.

[2] IPCC Climate Change 2014 Synthesis Report.

the industrial field is the main source of nitrous oxide emissions. In addition, the waste treatment process is also one of the sources of nitrous oxide emissions.

Industrial processes are the main source of the emissions of fluorine-containing gases. Controlling hydrofluorocarbon (HFC) emissions from industrial production processes is conducive to the coordinated management of the emissions of controlled substances under the Montreal Protocol, as well as to the promotion of the global application of HFC alternative technologies and products with a low global warming potential (GWP), and to the promotion of technologies to reduce emissions of perfluorocarbons (PFC) in aluminum electrolysis. The main application scenario for sulfur hexafluoride (SF_6) is in power supply equipment. Therefore, the recovery, treatment and reuse of SF_6 gas in power supply equipment should be strengthened. Nitrogen trifluoride is mainly used as an oxidant for rocket propellants, and also serves as an excellent etching gas in the plasma process of the microelectronics industry. It is also an excellent cleaning agent.

1.3 Time Attributes

In the framework of national goals, carbon neutrality is usually a concept applied to a period of one year, which means that annual carbon emissions need to be balanced by corresponding carbon sinks or carbon offsetting measures. This time attribute requires carbon emissions and carbon offsetting to be matched within the same period of time to ensure that annual carbon emissions do not produce a net increase in greenhouse gas concentrations in the atmosphere.

In order to encourage various regions, institutions, industries and organizations to actively participate in carbon neutrality practices, the policy allows the accumulation of emission reductions or carbon-sink capacity through emission reduction projects or carbon sink projects under the voluntary carbon emission reduction mechanism in the period prior to the target year for carbon neutrality. These projects may include afforestation and reforestation, carbon capture, utilization and storage technologies, and measures to improve energy efficiency, etc., which can generate emission reductions or carbon-sink capacity before the target year, and these accumulated carbon sinks can be used to offset future carbon emissions.

This approach provides organizations and regions with flexibility to plan and implement emission reduction measures according to their own circumstances while contributing to the long-term goal of carbon neutrality. In this way, a foundation for carbon neutrality practices can be gradually established, so that technological progress and innovation can be promoted, and experience and capabilities can be accumulated to achieve the ultimate goal of carbon neutrality.

2 Key Issues in Carbon Emissions Accounting

2.1 Selection of Base Period and Neutralization Period

Before implementing the measures for carbon neutrality, it is necessary to understand the carbon emissions of the target. At the regional level, it is necessary to analyze the industrial planning and status, functional area layout, the emission reduction level and the material flow circulation level, etc., to identify the gap with the advanced regional layout. At the enterprise and product levels, it is necessary to analyze the differences between the design, production and process, recycling and treatment processes and the current advanced low-carbon technologies in the industry, to find opportunities for improvement, and determine improvement measures.

For carbon neutrality projects at the project and activity level, it is necessary to identify their baseline scenarios and estimate the emission reductions of projects and activities. Therefore, carbon neutrality at the project and activity level does not have a base period, only a neutralization period.

Baseline periods and neutralization timelines should be a period of one year, a calendar year, or 365 days with the same start date. There can be a schedule for voluntary emission reduction between the baseline period and the neutralization period. The base period is usually the year before the carbon neutrality start point. If there is an imbalance in development or production, the previous three or five years can also be selected. In this case, the annual average should be used to reflect the carbon emissions. The neutralization period is usually implemented in one year.

2.2 Management of Sources and Sinks

The "source" of greenhouse gases refers to the process or activity that releases greenhouse gases into the atmosphere, while the "sink" of greenhouse gases refers to the process, activity or mechanism that removes greenhouse gases from the atmosphere.[3] The sources of greenhouse gases in the atmosphere are divided into natural sources and anthropogenic sources. The main reason for the increase in greenhouse gas concentrations in the atmosphere is the increase in anthropogenic sources caused by human activities.

In the process of implementing carbon verification and investigation during the base period, all emission sources and carbon sink projects within the scope of carbon neutrality should be identified. Based on the data availability and required precision, data sources with emissions no higher than 1% of their total can be omitted when appropriate, and this principle can also be applied to the treatment of carbon-sink capacity.

[3] United Nations Framework Convention on Climate Change.

2.3 Principle of Non-additionality

Under the voluntary carbon emission reduction mechanism, the existence of emission reductions is premised on whether the project has additionality, that is, such projects and their emission reductions, without the support of emission reduction mechanisms, have obstacles such as financial benefit indicators, financing channels, technical risks, market penetration and resource conditions, making it difficult to achieve normal operation based on the environmental conditions of the project owners. Once a project is proven to be able to operate normally without the support of emission reduction mechanisms, the project constitutes a baseline scenario and does not have additionality.

Under the carbon neutrality goal, the treatment of carbon sink projects does not need to consider the principle of additionality under the voluntary carbon emission reduction mechanism. It only needs to calculate the total amount of carbon sinks generated during the base period or reporting period within the accounting scope after deducting the emissions.

Once the emission reduction of an emission reduction project is included in other carbon neutrality implementation plans, it should not be included in the current carbon neutrality implementation plan. For example, if the electricity and emission reductions of a photovoltaic power generation project within the boundary of this carbon neutrality project are used by other projects outside the boundary, the net electricity or emission reductions of this photovoltaic project should not be deducted in this carbon verification.

3 Emission Reduction is Crucial to Achieving Carbon Neutrality

3.1 Voluntary Emission Reduction

Voluntary emission reduction is crucial to achieving carbon neutrality, and the emission reduction strategy is the core element of carbon neutrality. The carbon neutrality implementation plan has to include a voluntary emission reduction plan. This voluntary emission reduction plan should be based on a comparative analysis of the emission sources and carbon sinks identified in the baseline period in comparison with advanced industry practices or advanced technologies, and an appropriate diagnostic analysis should be conducted. Carbon neutral enterprises should formulate customized emission reduction plans. The carbon neutrality responsible parties at the project level should also formulate an emission reduction plan and implement as many feasible emission reduction measures as possible to ensure voluntary emission reductions at the project level.

At the regional level, emission reduction plans can be considered at the planning and operational levels, and comprehensive emission reduction plans can be formulated in the fields of energy, construction, transportation, agriculture, forestry, and industrial manufacturing. At the enterprise, project, and product levels, emission reduction targets can be achieved through low-carbon and zero-carbon technologies and corresponding emission reduction management.

At the national level, voluntary emission reductions can be achieved through formulating policies, establishing a system of standards, building basic capability, implementing plans, and designing guarantee mechanisms. These measures will lead the low-carbon energy revolution, build a low-carbon industrial system, promote low-carbon development in cities, and encourage some regions and industries to reach carbon peak first. In addition, it is necessary to carry out pilot demonstrations of carbon neutrality, enforce control over regional or industrial carbon intensity indicators, establish and operate a national carbon emissions trading mechanism, strengthen the research and development, demonstration and promotion of low-carbon technologies, etc.

In terms of energy, we should optimize the utilization of fossil energy, control the total amount of coal consumption, strengthen the clean and efficient utilization of coal, vigorously promote the substitution of transportation fuels with natural gas and electricity, and actively develop natural gas power generation and distributed energy. Carry out large-scale industrial demonstrations of carbon capture, utilization and storage in coal-based industries and oil and gas extraction industries, thus controlling carbon emissions in coal chemical and other industries. Actively develop and utilize natural gas, coalbed methane, and shale gas, and strengthen the recovery and utilization of vented natural gas and associated gases from oil fields. Accelerate the development of low-carbon energy, promote the planned development of hydropower, develop safe and efficient nuclear power, steadily develop wind power, accelerate the development of solar power generation, while actively developing geothermal energy, biomass energy and ocean energy. Adhere to the energy strategy of giving priority to energy conservation, rationally manage energy demand, and improve energy utilization efficiency. Strictly implement energy conservation assessment and review, and strengthen energy conservation supervision. Promote energy conservation and consumption reduction in key areas such as industry, construction, transportation, and public institutions. Implement the nation-wide energy conservation action plan and organize key energy conservation projects. Improve the energy conservation standard system, strengthen energy measurement supervision and services, and implement energy efficiency leadership actions. Promote contracted energy management and promote the healthy development of energy conservation service industry.

In terms of industrial system, green practices should be used as the foundation of development, which includes controlling emissions in the industrial sector, accelerating industrial structure adjustment, vigorously developing low-carbon agriculture, and increasing carbon sinks in ecosystems. Take low-carbon development as an important driving force for high-quality development, thus promoting the transformation and upgrading of the industrial structure. This requires the orderly elimination of backward and excess production capacity in accordance with laws and regulations, the application of advanced and applicable low-carbon technologies to transform traditional industries, the extension of the industrial chain, and the enhancement of the low-carbon competitiveness of enterprises. Reduce greenhouse gas emissions in the agricultural sector, which involves implementing zero-growth actions for fertilizer use, promoting formula fertilization based on soil test results, to reduce nitrous oxide emissions and control methane emissions from farmland. Select high-yield and low-emission crop varieties, and improve water and fertilizer management. Control greenhouse gas emissions from livestock and poultry farming, promote standardized and large-scale farming, and promote the comprehensive utilization of livestock and poultry waste. Build large and medium-sized biogas recovery and utilization projects in livestock and poultry farms according to local conditions. Furthermore, increase forest carbon sinks by implementing precise forest quality improvement projects, strengthening forest resource protection and disaster/hazard prevention, and controlling and reducing forest carbon emissions. Besides, enhance the carbon sequestration capacity of wetlands, actively increase the carbon sinks of grasslands, and explore pilot projects for carbon sinks in ecosystems such as oceans.

In terms of low-carbon development in cities and towns, implement low-carbon concepts and requirements in urban and rural planning, optimize urban functions and spatial layout, scientifically delineate urban development boundaries, explore new urbanization models that are intensive, intelligent, green, and low-carbon. Refine urban carbon emission management, improve infrastructure and construction quality, promote energy-saving renovation of existing buildings, strengthen energy-saving design and construction of new buildings, and achieve low-carbon operation and management of commercial and public buildings such as hotels, office buildings, and shopping malls. Promote low-carbon technologies such as waste heat utilization, high-efficiency heat pumps, renewable energy, distributed energy, green building materials, green lighting, and roof and wall greening according to local conditions. Carry out pilot demonstrations of carbon–neutral buildings. Build a low-carbon transportation system through accelerating the development of low-carbon transportation modes such as railways and water transportation, promoting low-carbon development of aviation, navigation, and road transportation, improving the urban transportation system that prioritizes public transportation, developing urban rail transit, intelligent transportation, and non-motorized transportation, and encouraging green travel. Use energy-saving, clean energy, and new energy transportation means, and improve supporting infrastructure construction. Promote the utilization of waste resources and low-carbon waste disposal. Advocate a low-carbon lifestyle.

3.2 *Carbon Offsetting*

Emission reductions or carbon sinks within the boundary of the carbon neutrality goal cannot be used for carbon offsetting. Emission reductions or carbon sinks outside the boundary may also have time attributes, if necessary, depending on the object for carbon neutrality.

Offset credits must adhere to the first-use principle, with blockchain-enabled registries ensuring permanent retirement to avoid double claiming. The carbon neutrality certification mechanism should establish an identification or traceability system that can identify the emission reductions or carbon sinks that have been used by this mechanism to facilitate query and identification.

In the early stages of carbon neutrality, in order to encourage enterprises to implement corresponding actions, they can choose the neutralization ratio during the carbon neutrality commitment stage and neutralize the committed carbon emissions ratio after implementing voluntary emission reduction measures. The recommended ratio is above 20%.

Chapter 3
Analysis of the Pathways to Achieving Regional Carbon Neutrality

Regional carbon neutrality is generally defined by geographic boundaries. The largest area is the entire Earth, but it is generally not possible for the Earth to achieve carbon neutrality due to challenges such as data collection and policy implementation. The second largest areas are the national or provincial/municipal regions for which the greenhouse gas emissions are calculated. Smaller geographic areas also include parks.

1 Characteristics of and Pathways to National Carbon Neutrality

1.1 Types of Greenhouse Gases

There are many types of greenhouse gases, and the greenhouse gases included the accounting in different countries and regions vary. The Kyoto Protocol specified the control of the following six greenhouse gases: carbon dioxide (CO_2), methane (CH_4), nitrous oxide (N_2O), hydrofluorocarbons (HFCs), perfluorocarbons (PFCs), and sulphur hexafluoride (SF_6). The latter three types of gases exhibit the highest global warming potential, but in terms of the percentage of contribution to global warming, CO_2 accounts for the largest proportion due to its high content. The IPCC Fifth Assessment Report added nitrogen trifluoride (NF_3), and many countries and regions have included NF_3 in their accounting scope.

At present, greenhouse gases are rarely calculated by direct measurement or in-line measurement. Instead, they are calculated using a specific formula based on the consumption of energy materials, emission factors and the corresponding coefficients. This calculation involves determining which types of greenhouse gases should be included. When China initiated the greenhouse gas pilot projects in two provinces

C. Chen et al., *Sustainable Carbon Neutrality*,
https://doi.org/10.1007/978-981-95-1691-9_3

and five cities in 2011, it initially calculated only CO_2. However, with the development of national carbon trading, different industries have included different types of greenhouse gases, and CO_2 is no longer the only one.

There are currently two probable scenarios. The first is a carbon neutrality goal that calculates only CO_2 emissions, not the other greenhouse gases. The second is a goal that calculates all the six greenhouse gases listed in the Kyoto Protocol, plus NF_3, which was added by the IPCC Fifth Assessment Report, a total of seven gases. At present, many countries in the world have clear target years for carbon neutrality, but the types of gases to be included have not yet been clarified. It is possible that some countries only include some gases other than CO_2. For example, New Zealand's largest source of emissions is agriculture. A law passed in November 2019 sets a net-zero target for all greenhouse gases except biogenic methane (mainly from sheep and cattle), which will be reduced by 24% to 47% from the 2017 level by 2050.

1.2 Methods of Making Carbon Neutrality Commitments

Carbon neutrality commitments at the national level can be made in two forms: legislation and policy, each with its own characteristics and functions. Legislative commitments are made to incorporate the carbon neutrality goal into the legal framework, thus ensuring its implementation and achievement through national coercion. This approach has the highest authority and binding force, providing long-term and stable targets and action guidelines for carbon neutrality. Legislation usually involves clear targets, timetables, responsible parties, enforcement mechanisms and supervision measures to ensure that the actions of the whole society are consistent with the national carbon neutrality goals. Policy commitments are action plans formulated by the government based on current conditions and goals. They are relatively flexible and can be adjusted according to actual conditions. Policies can serve as the basis for the formulation of laws to provide guidance and support for achieving the carbon neutrality goal. Although policies are not as binding as laws, they can clarify government intentions and priorities, influencing the behaviors of the public and private sectors. The two types of commitments complement each other. Laws provide a long-term and stable framework, while policies provide specific ways and means of achieving these goals. The flexibility of policies allows for rapid response to new challenges and opportunities, while the stability of laws ensures that the carbon neutrality goal remains consistent despite short-term political or economic fluctuations.

In practice, policies can serve as a vanguard to explore effective ways of reducing emissions through pilot and demonstration projects, thus providing experience and basis for legislation. Once the policy is verified and accepted, it can be turned into a binding law through the legislative procedures to ensure the long-term implementation and realization of the carbon neutrality goal. In short, both legislative commitments and policy commitments are important means for a country to achieve its carbon neutrality goals. The key lies in the appropriate use of these two methods according to the specific conditions and needs of the country, ensuring that the realization of the carbon neutrality goal has both a solid legal basis and flexible policy adaptability.

1.3 Decomposition and Refinement of Goals

Countries that have set carbon neutrality goals have basically set the target years as 2050 and 2060. To achieve the carbon neutrality goal, especially between 2050 and 2060, it is necessary to carefully decompose and refine this ambitious goal and design a clear path to achieve it. This process involves breaking down long-term goals into short- and medium-term ones, ensuring that each stage has clear targets and action plans.

First, it is necessary to comprehensively plan and adjust the energy structure and clarify the scale and proportion of renewable energy such as solar energy, wind energy, and hydrogen energy. This includes determining the number of photovoltaic power plants and wind power generation facilities to be built in specific areas, as well as the layout and promotion of hydrogen energy and other clean energy technologies. Second, it is necessary to formulate specific industrial emission reduction strategies, including energy efficiency improvement and low-carbon transformation in sectors such as industry, transportation, construction, and agriculture. This involves technological innovation, adjustment of industrial structure, and changes in energy consumption patterns. Finally, it is also necessary to consider the enhancement of carbon sinks, such as improving the carbon absorption capacity of natural systems through ecological projects such as afforestation and wetland protection. At the same time, carbon capture, utilization, and storage (CCUS) technologies should be explored and implemented to address missions that are difficult to reduce.

China plans to complete the transition from carbon peaking to carbon neutrality in about 30 years, which is more urgent and more challenging than the 50-year plan in some developed countries in Europe and America. This requires China to take more proactive and efficient measures in policy formulation, technological innovation, capital investment and international cooperation. A detailed roadmap will facilitate smoother implementation, thereby increasing the possibility of achieving the goal. This requires the joint efforts of the government, enterprises and all sectors of society to ensure the smooth realization of the carbon neutrality goal through scientific planning, well-designed layout and effective implementation.

1.4 Characteristics of Regional Carbon Neutrality

Achieving carbon neutrality at the regional level is a complex and comprehensive process, which has the following notable characteristics.

a. Deep Decoupling Between Economic Development and Carbon Emissions

Achieving carbon neutrality at the regional level is a comprehensive and far-reaching goal that requires a deep decoupling between economic development and carbon emissions. This decoupling means that, although the scale of the economy may continue to grow, carbon emissions are effectively controlled through technological innovation, policy guidance and market mechanisms, and no longer positively correlate with economic growth. In carbon-neutral regions, the energy structure will undergo a fundamental transformation, dominated by renewable energy and clean energy sources. At the same time, the improvement of energy efficiency and the widespread application of low-carbon technologies will significantly reduce carbon emissions. In addition, the transformation of social awareness, the public's acceptance and practice of low-carbon lifestyles, and the development of low-carbon industries are all indispensable components for achieving carbon neutrality. This requires the joint efforts of governments, enterprises and individuals to build a sustainable low-carbon economic system through capital investment, technological research and development, and policy support.

b. Low-Carbon Development of Energy Resources

In a carbon-neutral region, low-carbon development of energy sources is the key to achieving this goal. This involves a fundamental adjustment to the energy structure, in which the use of traditional fossil energy, such as coal and oil, will be significantly reduced. Instead, renewable energy will be widely used, including solar energy, wind energy, geothermal energy, ocean energy and bioenergy, etc. The development and use of these energy sources will provide society with a clean and sustainable energy supply. At the same time, nuclear energy, as an efficient and clean energy source, will also be fully utilized under the premise of ensuring safety to meet the growing energy demand.

Low-carbon development of energy sources also includes improving the efficiency of energy use, reducing energy consumption through technological innovation and system optimization, and developing and applying advanced technologies such as smart grids to manage and distribute energy in a more efficient and environmentally friendly way. In addition, low-carbon development of energy sources also means transforming and upgrading existing energy systems, as well as constantly researching, developing and investing in emerging energy technologies to ensure the reliability of energy supply and environmental sustainability.

Achieving low-carbon development of energy sources is a long-term and systematic process that requires the joint efforts of policymakers, energy companies, scientific research institutions and all sectors of society. By formulating reasonable policies, promoting technological innovation, strengthening international cooperation and raising public awareness, we can jointly promote energy transformation and lay a solid foundation for achieving the carbon neutrality goal.

c. Regional Carbon Emissions Reach Zero or Negative Emissions

Achieving zero or negative regional carbon emissions is the ultimate manifestation of the carbon neutrality goal, which requires that the carbon dioxide generated from human activities in this region can be effectively absorbed or converted by natural processes and technological innovation. These include the use of plant photosynthesis to absorb carbon dioxide from the atmosphere, and the natural absorption of carbon dioxide by the ocean. In addition, negative emission technologies, such as carbon capture and storage (CCS), carbon capture, utilization and storage (CCUS), and enhanced rock weathering, are key technical means to achieve negative emissions. These technologies can remove carbon dioxide directly from the atmosphere or store it underground for a long time, thereby reducing the concentration of greenhouse gases in the atmosphere.

In order to achieve zero or negative emissions, the region needs to take a set of comprehensive measures, including but not limited to improving energy efficiency, promoting low-carbon technologies and renewable energy, developing green transportation systems, improving land use and forest management, etc. These measures need to be combined with the region's economic development, social needs and environmental protection objectives to form a coordinated sustainable development strategy.

In addition, achieving zero or negative regional carbon emissions also requires policy support, capital investment, technological innovation and public participation. The government can guide enterprises and individuals to take low-carbon actions by formulating relevant policies and incentives, while increasing investment in the R&D and application of low-carbon technologies to promote the formation of a low-carbon economic and social development model. The public's environmental awareness and behavior changes are also important for achieving this goal. Education and publicity are needed to raise public awareness of climate change and encourage everyone to adopt lifestyles that save energy and reduce emissions.

In short, achieving zero or negative regional carbon emissions is a multi-dimensional, cross-domain and systematic undertaking that requires the joint efforts and collaboration of the government, enterprises, scientific research institutions and the public. Through scientific and technological innovation, policy guidance and changes in behaviour, we can jointly advance the region towards carbon neutrality.

d. The Low-Carbon Concept Is Strongly Supported by the People

The strong support for the low-carbon concept by the people is the social foundation for achieving the goal of carbon neutrality. The wide acceptance of this concept means that the public's awareness and actions regarding resource conservation and environment protection will reach a new level. In a carbon–neutral society, low-carbon living is not just a choice, it is also part of people's daily lives, reflected in everyone's behavior and consumption patterns.

This transformation is reflected in many aspects: from saving electricity and water at home to making green choices in daily travel; from reducing food waste to preferring sustainable products and services; from supporting environmentally friendly products to actively participating in community and public activities to protect the environment. The popularization of low-carbon concepts has also raised the public's awareness of climate change issues and brought higher expectations and demands on policymakers and companies to take action in environmental protection.

Education and the media play an important role in promoting low-carbon concepts. Through school education, publicity, social media and other channels, the knowledge, skills and values associated with low-carbon living are disseminated and strengthened. Enterprises and social organizations can also encourage and guide the public to participate in the practicing of a low-carbon lifestyle by providing low-carbon products and services and carrying out environmental protection activities.

In addition, policymakers have provided support and guarantees for the popularization of low-carbon concepts through legislation and incentives, such as carbon taxes, subsidies, and green credits. These policies not only encourage companies and individuals to reduce carbon emissions, but also reflect the society's emphasis on sustainable development.

In short, the key to achieving the goal of carbon neutrality is the acceptance of low-carbon concepts by the people. It requires joint efforts from all sectors of society to foster low-carbon lifestyles and consumption patterns throughout society through education, publicity, policy guidance and market mechanisms. Such a social transformation will provide a solid social and cultural foundation for achieving carbon neutrality.

e. Rapid Development of Low-Carbon Technologies

The rapid development of low-carbon technologies is an indispensable driving force for achieving the goal of carbon neutrality. These technologies cover all aspects from energy supply to demand-side management, including but not limited to smart grids, solar energy, wind energy and other renewable energy technologies that can effectively replace traditional fossil fuels and reduce greenhouse gas emissions. At the same time, energy efficiency improvement technologies on the demand side, such as building energy conservation, industrial process optimization, and transportation system improvements, are also crucial to reducing energy consumption and carbon emissions.

Carbon capture, utilization and storage (CCUS) technology is a means of controlling carbon dioxide emissions at the end of the process, which is of great significance for achieving carbon neutrality. CCUS technology can capture carbon dioxide from industrial emission sources or the atmosphere and convert it into useful products or store it safely underground, thereby reducing the concentration of greenhouse gases in the atmosphere.

In addition, the rapid development of low-carbon technologies also includes the improvement and innovation of existing technologies, such as improving energy conversion efficiency, developing new materials, optimizing energy management systems, etc. The development of these technologies can not only improve energy efficiency and reduce costs, but also promote the development of new business models and industries.

Policy support, capital investment, R&D cooperation and the improvement of market mechanisms are all important for promoting the development of low-carbon technologies. The government can encourage enterprises and research institutions to increase their investment in low-carbon technologies by formulating relevant policies, providing R&D funds, establishing technical standards and certification systems, etc. At the same time, international cooperation and technical exchanges can also accelerate the innovation and application of low-carbon technologies.

The rapid development of low-carbon technologies is an important way to achieve the goal of carbon neutrality. It requires the joint efforts and collaboration from all sectors of society. We can jointly promote the widespread application of low-carbon technologies through technological innovation, policy guidance and market mechanisms, thus providing solid technical support for building a low-carbon society.

f. Unprecedented Development of Low-Carbon Industries

The unprecedented development of low-carbon industries is a key component in achieving the goal of carbon neutrality, which involves many aspects such as technological innovation, industrial transformation, market mechanisms and the policy environment. These comprehensive measures can drive the economy towards low-carbon, environmentally friendly and sustainable development, thus laying a solid industrial foundation for responding to global climate change. Achieving the carbon neutrality goal will drive the unprecedented development of low-carbon industries. As society's demand for low-carbon technologies and products increases, the industrial structure will be optimized and upgraded to adapt to this new economic model. The development and utilization of renewable energy sources in the low-carbon energy field, such as solar energy, wind energy and hydropower, will be expanded on a large scale and become the dominant force in energy supply.

Low-carbon industries, including low-carbon materials, green buildings, electric vehicles and energy storage technologies, will witness technological innovation and market expansion. Low-carbon agriculture improves the environmental sustainability of food production through sustainable farming methods and practices that reduce agricultural emissions. Low-carbon manufacturing reduces energy consumption and waste generation in industrial production processes by adopting clean production technologies and circular economy principles.

In addition, comprehensive one-stop energy-saving services will become an important player in supporting enterprises to reduce energy consumption and carbon emissions. These services may include energy audits, the design and implementation of energy efficiency improvement solutions, and ongoing energy management support. Low-carbon financial services, such as green bonds, green funds, and carbon trading markets, will provide the necessary financial support for low-carbon projects, promoting the research, development and application of low-carbon technologies.

As the goal of carbon neutrality becomes more accepted, relevant education, training and professional consulting services will also be developed to meet market demand for low-carbon knowledge and expertise. At the same time, policymakers will create favorable conditions for the development of low-carbon industries by formulating incentive measures and regulations.

In general, regional carbon neutrality is a systematic undertaking involving multiple aspects such as the economy, energy, technology, and society. It requires us to maintain economic growth while achieving effective control and reduction of carbon emissions through technological innovation, policy guidance, and public participation, thereby building a sustainable and low-carbon society.

1.5 Pathways to Achieving the Goal

Achieving the carbon neutrality goal requires a series of comprehensive measures, including deep decarbonization of the energy system, replacing high-carbon energy with low-carbon energy by accelerating the development of renewable energy and building intelligent power systems; at the same time, improving energy efficiency by strictly controlling energy consumption and promoting energy-saving technologies, especially in the fields of information infrastructure such as 5G networks and artificial intelligence; in addition, developing and deploying low-carbon technologies such as energy storage, hydrogen energy, and carbon capture, utilization and storage technologies, promoting the integration of these technologies with digital technologies, and implementing them in high-energy-consuming industries such as steel and cement; finally, increasing ecological carbon sinks and enhancing the carbon sequestration capacity of natural ecosystems through measures such as land greening, forest resource management and ecological restoration. These measures together constitute a comprehensive strategy for achieving the carbon neutrality goal, and countries are studying and formulating specific implementation roadmaps based on their specific context.

Achieving carbon neutrality at the regional level is a multi-dimensional, cross-domain and systematic undertaking, which requires the adherence to the following implementation pathways.

a. Clarify the Top-Level Design

Achieving carbon neutrality at the regional level requires a comprehensive and coordinated design at the top level, which involves formulating a clear development strategy and goals, establishing a practical timetable and roadmap, and ensuring coordination with social and economic development. This process requires the in-depth participation from industries and industry experts, who will carry out scientific planning of low-carbon energy, low-carbon industrial transformation, improvement of energy efficiency, and establishment of institutional and standard systems based on the specific conditions of the region. At the same time, it is also necessary to build a comprehensive communication platform to discuss and deploy key tasks and key projects for carbon neutrality, and clarify the responsibilities and obligations at all levels of governments, enterprises and the public.

In addition, a series of supporting policies and legal documents need to be adopted to establish a guarantee mechanism and raise the public's awareness of the importance of carbon neutrality, thus unifying the mindset and actions. By consolidating the main responsibilities of various industries and local governments, a consistent policy system can be formed to ensure that the carbon peaking and carbon neutrality goals are achieved as scheduled. This includes developing low-carbon finance, directing capital flows to low-carbon projects, supporting technological innovation and industrial upgrading, and establishing a carbon emission monitoring, reporting and verification mechanism to ensure the effective implementation of carbon neutrality measures.

b. Basic Capacity Building

Basic capacity building is the premise and support for achieving regional carbon neutrality goals. First, it is necessary to strengthen the monitoring, reporting and verification system of carbon emissions to ensure the accuracy and transparency of data and to provide the necessary data support for policy formulation and implementation. As the process towards carbon neutrality progresses, the infrastructure transformation has become the key. Power facilities will develop from relying on high-emission fossil fuels to relying mainly on low-emission renewable energy, supplemented by natural gas and hydrogen energy storage. This transformation is accompanied by the natural and, in some cases, forced decommissioning of existing infrastructure. At the same time, the infrastructure for emerging industries such as wind power, photovoltaics, terminal consumption of power, energy storage, and carbon capture, utilization and storage will also be improved and expanded.

The power grid system is facing the complexity brought about by the large-scale and unstable connection of distributed renewable energy to the power grid, which requires the power grid to improve its flexibility and scalability. The construction of grid interconnection and smart grid has become particularly important, as they will support the energy structure dominated by renewable energy and complemented by hydrogen energy. Combined with energy storage technology, they help meet the growing demand for energy while ensuring the affordability, reliability and sustainability of energy supply.

In addition, when building and upgrading infrastructure, the maturity and cost-effectiveness of technologies need to be taken into account to ensure the economic feasibility and environmental benefits of the transformation process. The synergy among policy support, technological innovation and market incentive mechanisms is essential to facilitate this process. These measures can provide a solid foundation for achieving the carbon neutrality goals and ensuring the sustainable development of the regional economy and society.

c. Establishing a System of Policies and Standards, and Publishing Carbon Neutrality-Related Systems

Achieving regional carbon neutrality requires the establishment of a comprehensive system of policies and standards, which is a key framework to ensure the realization of the carbon neutrality goal. This includes formulating a set of policies and standards related to carbon neutrality, such as carbon pricing mechanisms, emission standards, and energy efficiency standards. These policies and standards will guide and regulate the behavior of various industries and enterprises, thus promoting a low-carbon transformation.

Policy formulation must ensure the establishment of laws and regulations to guide carbon neutrality actions. Implement the leadership accountability system, and strictly hold accountable administrative departments and relevant leading cadres or administrative personnel who perform poorly in carbon neutrality construction, to ensure that various policies and measures are effectively implemented. In addition, we should provide diversified and improve carbon trading mechanisms at different levels, promote the development of multi-level and multi-sector carbon markets, thus making carbon trading an effective supplementary mechanism for carbon emissions. The carbon trading mechanism can provide economic incentives for emission reductions and encourage enterprises to take more proactive measures to reduce emissions.

It is crucial to build a complete system of financial mechanisms for carbon neutrality. This includes establishing a pricing system and price setting mechanism that covers carbon elements, ensuring that the costs and benefits of carbon emissions are properly reflected, and making carbon an economic consideration as important as other production factors. At the same time, it is necessary to improve the operating mechanism of the carbon financial market, reduce operating costs, and improve efficiency. Building a multi-level and wide-ranging carbon financial system can provide financial support for carbon neutrality projects and promote the development of low-carbon technologies and industries.

The formulation and implementation of these policies and standards can provide solid policy support and market mechanisms for regional carbon neutrality, ensuring that economic and social development aligns with ecological and environmental protection, and achieving the goal of sustainable development.

d. Direct Emission Reductions

Direct emission reduction is an important part of achieving the carbon neutrality goal and covers several key aspects, such as improving energy efficiency, reducing the use of carbon-containing energy sources and continuous energy conservation.

Firstly, technological progress is crucial in all aspects of power generation, transmission, distribution and energy use. This includes adopting more efficient power generation technologies, improving transmission and distribution networks to reduce the loss, and adopting energy-saving equipment and intelligent management systems in end-use applications. These technological improvements help to improve the efficiency of energy use, reduce energy consumption, and thus reduce carbon emissions.

Secondly, reducing the use of carbon-containing energy is the core of achieving carbon neutrality. This requires us to gradually reduce our dependence on fossil fuels such as coal, oil and natural gas, and switch to renewable energy sources such as wind and solar energy, and carbon-free energy sources such as hydrogen energy. These clean energy sources can not only provide a continuous and stable supply of energy, but also achieve zero-carbon emissions during their use.

Finally, constant energy-saving measures are also essential. This requires raising public awareness of low-carbon energy conservation through policy publicity and education, and encouraging people to take energy-saving actions in their daily lives to reduce energy waste. The government can promote the application and popularization of energy-saving technologies by formulating energy-saving standards, and providing tax incentives and subsidies.

Achieving carbon neutrality requires a fundamental shift in the energy structure, transitioning from a reliance on carbon-containing energy to an energy system dominated by carbon-free and low-carbon energy. This is not only a technical issue, but also a socio-economic issue that requires the joint efforts and participation from the government, enterprises and the public. Through the implementation of direct emission reduction measures, we can effectively reduce carbon emissions, promote the sustainable development of the economy and society, and ultimately achieve the goal of carbon neutrality.

(i) Continuous Low-Carbon Development and Conversion to Electricity of Energy Sources

Continuous low-carbon development and conversion to electricity of energy sources is one of the key paths to achieve regional carbon neutrality. This transformation requires the vigorous development of renewable energy, the gradual reduction of dependence on fossil energy, and the increase of the proportion of electricity in terminal energy consumption. At present, the proportion of electricity in China's terminal energy consumption is about 25%, but according to the 1.5 °C temperature control target and the vision of achieving carbon neutrality by 2060, this proportion needs to be increased to about 70% by 2050.

With the progress in achieving carbon neutrality, electricity will replace the direct combustion of fossil energy on a large scale, which involves not only increasing the proportion of primary energy converted into electrical energy, but also strengthening the clean and low-carbon attributes of electricity. The constant low-carbon development of energy means that the energy industry needs to undergo profound structural adjustments. The proportion of fossil fuels as basic energy supply will be greatly reduced and replaced by clean energy such as wind power, solar energy, and biomass energy. In the United States' carbon neutrality action plan, coal is expected to be completely replaced, and the use of natural gas and oil will also be greatly reduced by 75% and 90% respectively. Natural gas and oil will mainly be used as industrial raw materials for specific modes of transportation, and as a limited source of energy for natural gas-fired power plants that play a role in maintaining electricity stability. This shows that electricity will become the main form of terminal energy consumption, and the energy conversion process will become more important.

Advanced biofuel refining technologies, as well as the production of hydrogen and synthetic fuels through electrolysis, will become key components in building a carbon–neutral energy system. The development and application of these technologies will help build a cleaner, more efficient and sustainable energy system.

To achieve this transformation, policy support, technological innovation, infrastructure development, market mechanisms and public awareness need to be promoted all together. Policymakers need to introduce corresponding incentives and regulatory standards to foster the development and application of clean energy; scientific research institutions and enterprises need to intensify their R&D efforts to drive the advancement of energy technologies; infrastructure construction needs to keep up with the time to adapt to the development needs of new energy sources; market mechanisms need to be improved to enhance the cost-effectiveness and accessibility of clean energy; and the improvement of public awareness will help to create a social atmosphere that supports a low-carbon lifestyle. Through these comprehensive measures, the energy system at the regional level will be able to achieve a fundamental low-carbon transformation and lay a solid foundation for achieving the carbon neutrality goal.

(ii) Continue to Save Energy and Improve Energy Efficiency

Continuously saving energy and improving energy efficiency is one of the key measures to achieve the goal of carbon neutrality. These measures involves not only technological innovation and management optimization, but also reducing energy consumption and improving energy efficiency in various sectors and at different levels, thereby reducing energy consumption per unit of GDP.

Since renewable energy sources such as wind, solar, and biomass are constrained by time and geographical locations, improved energy efficiency and constant energy conservation are essential to reducing the dependence on these resources. This helps to reduce the pressure to further expand renewable energy generation capacity to meet energy demand. Energy efficiency improvements and energy conservation measures themselves are important components of the decarbonization strategy, contributing to achieving carbon neutrality at the end of energy production and consumption. They can be regarded as a supplement to renewable power generation capacity, and are indispensable for achieving the carbon neutrality goal.

The dual energy control strategy, controlling total energy consumption and energy intensity, will continue to play a role in the long-term process of carbon neutrality. This requires improving energy efficiency and saving energy in various sectors, including energy, industry, transportation, construction, agriculture and social life. For example:

In the industrial field, advanced manufacturing processes and equipment are adopted to optimize production processes and reduce energy waste.

In the transportation sector, promote electric vehicles and public transportation systems, and improve fuel efficiency and emission standards.

In the construction sector, implement energy-saving design and green building standards, and adopt energy-saving materials and technologies.

In the agricultural sector, energy-saving irrigation and farming technologies are adopted to improve the efficiency of agricultural production.

In social life, the public's awareness of energy conservation is raised and the use of energy-saving appliances and equipment is encouraged.

In addition, policymakers need to introduce appropriate incentives and regulations, such as energy efficiency labelling, tax incentives, and subsidy policies, to promote the research and development and application of energy-saving technologies. At the same time, training and education in energy management and energy-saving technologies should be strengthened, so as to improve the energy-saving awareness and capabilities of the public and enterprises. Through these comprehensive measures, energy consumption can be effectively reduced, energy efficiency can be improved, and a positive contribution can be made to achieving the goal of carbon neutrality.

e. Promote Selected Regions and Industries to Take the Lead in Achieving Carbon Neutrality

Encouraging selected regions and industries to take the lead in achieving carbon neutrality is a phased, differentiated strategy that aims to provide a reference model for other regions and industries to achieve carbon neutrality through the experience of success of pilot regions. This strategy recognizes the differences in resource availability, industrial bases, industrial layouts and levels of economic development among different regions, thereby allowing and encouraging those with mature conditions to take the lead.

Achieving a nationwide carbon neutrality target by 2060 requires a flexible timetable that allows different regions to develop appropriate carbon peaking and neutrality roadmaps based on their own circumstances. Some regions may achieve carbon neutrality earlier because they have abundant renewable energy sources, advanced technologies and high levels of economic development. These regions can serve as pilot areas to explore effective emission reduction technologies and management models in practice, thus generating experiences and practices that can be replicated and popularized.

For those energy-intensive industries, cities and regions with heavy industries, a reasonable timeframe can be set to allow them to gradually achieve carbon neutrality according to their own transformation needs and capabilities. At the same time, for those industries that are crucial to economic and social development, some flexibility can be given to allow them to achieve carbon neutrality on a later date, while ensuring that the overall carbon neutrality goal is not compromised.

This differentiated strategy requires the government to set precise policies that ensure the realization of the overall goal while taking into account the particularities of regions and industries. By providing policy support, technical guidance and financial incentives, we can encourage and guide qualified regions and industries to take the lead in action taking, thus driving the carbon neutrality process of other regions and industries through demonstration effects.

In addition, this strategy also requires the establishment of a fair evaluation and incentive mechanism to ensure that all regions and industries can obtain the necessary support in the process of achieving carbon neutrality. At the same time, recognition and rewards should be given to those regions and industries that have achieved remarkable results, so as to create positive incentives and encourage the whole society to participate in the great cause of carbon neutrality. In this way, the national carbon neutrality goal can be gradually achieved, and a clean, low-carbon and sustainable future development model can be built.

f. Improving Ecological Carbon Sink Capacity

Carbon sinks are natural solutions to achieving carbon neutrality goals, involving the absorption and storage of the carbon dioxide from the atmosphere through the processes and activities within the natural ecosystems. These activities mainly include afforestation, forest management, vegetation restoration, etc. Through these measures, plants absorb carbon dioxide through photosynthesis, converting it into organic matter and sequestering it in plant biomass and soil for a long period of time. The role of carbon sinks is to enhance the Earth's carbon absorption capacity, thereby helping to offset carbon emissions generated by human activities. They not only contribute to reducing greenhouse gas concentrations in the atmosphere, but also have a positive impact on protecting biodiversity, maintaining ecological balance and improving environmental quality. In order to effectively utilize carbon sinks, a series of measures need to be taken, including rational plans for the planting of forests and other vegetation planting, strengthening the protection and management of forest resources, and restoring degraded lands and ecosystems. In addition, scientific research and technological development can also improve the efficiency and sustainability of carbon sinks, such as by selecting tree species and vegetation types with high carbon sequestration capacities, and by adopting advanced forest management technologies.

Improving ecological carbon sink capacity is one of the important ways of achieving the goal of carbon neutrality. As the largest terrestrial and marine carbon sinks on Earth, natural ecosystems can absorb and store large amounts of carbon dioxide through natural processes such as photosynthesis. By taking a series of measures, such as afforestation, wetland protection, and ecological restoration, the carbon sink function of these systems can be effectively enhanced and the regional carbon absorption capacity can be improved. Carbon neutrality does not mean achieving zero carbon emissions, because even under the most stringent emission reduction scenario, some industries and processes may still emit a certain amount of carbon dioxide. In this case, the role of carbon sinks is particularly important. For example, Liu Lianbo, an expert from China Huaneng Group, pointed out that even if extreme emission reduction measures are implemented in the power industry, there may still be about 1.5 billion tonnes of carbon dioxide emissions by around 2050. In order to achieve carbon neutrality, it is necessary to rely on carbon sinks and other technical means to offset these remaining emissions.

Types of carbon sinks include, but are not limited to, natural and semi-natural systems such as forests, agricultural land, and oceans. These carbon sinks can not only absorb carbon dioxide through natural growth processes, but can also enhance their carbon sink capacity through appropriate land management and ecological engineering. However, it should be noted that the exploitable carbon sink resources are finite, and the amount of carbon dioxide absorbed by natural carbon sinks is currently between 700 and 800 million tonnes per year.

In order to more effectively utilize and enhance carbon sink capacity, measures that can be taken include afforestation, wetland protection and restoration, ecological restoration, agricultural management, ocean carbon sinks, etc., as well as carbon sink monitoring and evaluation, formulation of supportive policies and incentives, and international cooperation. These measures can maximize the carbon sink potential of natural ecosystems, contributing to the realization of carbon neutrality goals. At the same time, this will also help protect biodiversity, improve the resilience of ecosystems, and mitigate the impact of climate change.

g. Carbon Dioxide Capture, Utilization and Storage (CCUS)

Carbon Capture, Utilization and Storage (CCUS) technologies are crucial to achieving the goal of carbon neutrality. It provides an effective solution for industries where emission reductions are challenging through other means. CCUS captures carbon dioxide through technical means, and uses or safely stores it to prevent it from being released into the atmosphere, thereby reducing the concentration of greenhouse gases. Although carbon neutrality does not mean no carbon emissions at all, CCUS technologies can help us neutralize these emissions through alternative means and achieve net zero emissions. The assessment report of the Intergovernmental Panel on Climate Change (IPCC) of the United Nations also emphasized the importance of CCUS technologies, pointing out that without CCUS, most climate models cannot achieve emission reduction targets and the cost of emission reduction will increase significantly. CCUS is not only a key strategic technology for achieving carbon neutrality, but also the primary technical pathway to achieve large-scale low-carbon development of fossil energy. In order to promote the development and application of CCUS technologies, policy support, technological innovation, international cooperation and constant capital investment are needed. Through these measures, CCUS technologies will play a more important role in global emission reduction efforts, contributing to building a clean, low-carbon and sustainable future.

Table 1 compares the difficulty, cost, upper limit and scope of different pathways to carbon neutrality, including direct emission reductions, offsetting, carbon sinks, CCUS, etc.

Achieving regional carbon neutrality requires the joint efforts of the government, enterprises, the public and the international community. Through various means such as scientific and technological innovation, policy guidance, market incentives, and international cooperation, a carbon neutrality action system involving the whole society can be established. At the same time, it is also necessary to constantly monitor and evaluate the progress towards carbon neutrality, and to timely adjust and optimize the implementation strategy to ensure the smooth realization of the carbon neutrality goal.

Table 1 Comparison between pathways to carbon neutrality

Method	Difficulty	Cost	Upper limit	Scope of use	Constraints
Direct emission reductions	Easy at first, then hard	Gradually increase	Yes	Wide	Unlimited emission reductions are impossible
Offsetting	Easy	Low	Yes	Limited	Policy
Carbon sinks	Easy	Low	Yes	Limited	Land
CCUS	Easy	High at first, then progressively lower	No	Limited	Degree of carbon dioxide dispersion

h. Offsetting Mechanism

As a complementary strategy, carbon offsetting plays an vital role in neutralizing carbon emissions. It is typically achieved by tree planting, energy-saving initiatives, or purchasing emission reduction allowances, thereby helping individuals or companies balance their unavoidable carbon emissions. Carbon offsetting is applied in multiple agreements and markets around the world, such as the relevant mechanisms outlined in the Kyoto Protocol and China's carbon trading market. Although many countries have announced carbon neutrality goals, the specific pathways to achieving these goals are still being explored. Different countries may develop their own offsetting mechanisms based on their development level, resource availability and policy environment. Some countries, such as Sweden, have clearly defined the strategy of achieving most of the emission reductions through domestic policies and the remaining part through international cooperation. Table 2 lists the carbon offsetting mechanisms that are being implemented or approved. These mechanisms have their own advantages and disadvantages, and it is necessary to choose the appropriate mechanism according to the specific context to achieve the emission reduction goal.

The effectiveness and sustainability of the carbon offsetting mechanism depends on a clear definition of the offsetting carbon sources, including determining the types of emission reduction projects that can be used for offsetting and how to ensure the long-term and environmental integrity of these projects. At the same time, international cooperation plays a key role in carbon offsetting, requiring the consideration of whether carbon emission reductions in different countries and regions can be used for offsetting, and how to establish a fair and transparent international carbon trading market. In addition, the carbon offsetting mechanism requires strict supervision and transparency to prevent fraud and double counting problems. Technological innovation and financial support are also key factors that will help reduce the cost of carbon offsetting projects and increase their feasibility. As countries gradually clarify their roadmaps to carbon neutrality, the carbon offsetting mechanism will become more specific and standardized, becoming an important tool for achieving global carbon neutrality goals.

It should be emphasized that offsetting is not a panacea, nor is it the only way. The regional approach to carbon neutrality should be science-based. Firstly, we need to determine the carbon neutrality approach, fully implement direct emission reductions, and promote a series of measures such as carbon capture, utilization and storage to minimize the carbon footprint of the subject. Only when these measures are taken, can offsetting be considered for the remaining carbon emissions that cannot be eliminated. The regional approach to carbon neutrality should also be adjusted according to actual conditions and needs, taking into full account of the specific characteristics and context of the regional subject.

Table 2 Carbon offsetting mechanism

Climate policy	Mechanism	Content	Emission reduction
Kyoto protocol	Clean development mechanism	The Clean Development Mechanism allows Annex I Parties to the United Nations Framework Convention on Climate Change to jointly carry out greenhouse gas emission reduction projects such as carbon dioxide with non-Annex I Parties. The emission reductions generated by these projects can be used by Annex I Parties[1] to fulfill their commitments to limit or reduce emissions. For developed countries, CDM provides a flexible mechanism for compliance; for developing countries, through CDM projects, they can obtain partial financial assistance and advanced technologies, which is conducive to developing countries' ultimate realization of the goals of the United Nations Framework Agreement on Climate Change	CERs
	Joint implementation mechanism	The Joint Implementation Mechanism is a project-based cooperation mechanism among Annex I Parties to the United Nations Framework Convention on Climate Change. Its purpose is to help Annex I Parties achieve their quantified greenhouse gas emission reduction commitments at a lower cost. Annex I Parties with higher emission reduction costs implement greenhouse gas emission reduction projects in Annex I Parties with lower emission reduction costs through this mechanism	ERUs
	International emissions trading scheme	The international emissions trading mechanism allows the parties to Annex I of the United Nations Framework Convention on Climate Change to trade carbon emission allowances with each other to reduce the negative impact of greenhouse gas emission reduction activities on the economy and achieve the most cost-effective global emission reduction. Its core is that developed countries can trade their surplus emissions within the quantitative limits stipulated in the Kyoto Protocol	EUAs

(continued)

[1] Annex I Parties/Non-Annex I Parties: The United Nations Framework Convention on Climate Change (hereinafter referred to as the Convention) was adopted at the United Nations Conference on Environment and Development in 1992. It is the basic framework for the international community to cooperate in addressing global climate change. Currently, about 200 countries or regions have joined the Convention as its parties. The parties are divided into Annex I Parties and Non-Annex I Parties. Non-Annex I Parties include developing countries, while Annex I Parties are developed countries. The two parties bear different responsibilities.

Table 2 (continued)

Climate policy	Mechanism	Content	Emission reduction
Voluntary emission reduction	Gold Standard	Gold Standard is a quality standard for Clean Development Mechanism and Joint Implementation projects, providing the first independent and optimal implementation standard for emission reduction projects under the Clean Development Mechanism (CDM) and Joint Implementation (JI). The standard can be used as a tool for project implementers to ensure the environmental benefits of their projects, which are equivalent to additional investments in sustainable energy services	GS VERs
Voluntary emission reduction	Verified Carbon Standard	Reductions can be made by logging into the platform and applying for the issuance of verified carbon units (VCU) reduction allowances from the international Verified Carbon Standard (VCS). VCS is currently the most widely used voluntary greenhouse gas emission reduction program in the world. VCS allows certified projects to convert their greenhouse gas emission reductions and removals into tradable carbon credits	VERs
China's carbon trading mechanism	China Certified Emission Reduction	For the voluntary emission reductions generated after the government approves the filing, key emission generation organizations can use a certain proportion of China Certified Emission Reductions (CCERs) that meet the requirements to fulfill the reduction requested	CCER

2 Practices of Achieving Carbon Neutrality in Some Countries

The consequences of global warming include melting polar glaciers, rising sea levels, frequent extreme weather events, land desertification, and ocean acidification. The United Nations' report pointed out that the number of global climate disasters surged between 2000 and 2019, an increase of 83% over the previous 20 years. In response to these challenges, many developed countries have set a timetable for carbon neutrality. Finland, for example, plans to achieve net-zero emissions by 2035, while Sweden and Austria have set 2045 as their target. Countries and regions such as the European Union, the United Kingdom, Norway, Canada and Japan have set the goal for 2050. Some developing countries, such as Chile, have also proposed plans to achieve carbon neutrality by 2050. In September 2020, China announced an ambitious goal of reaching carbon peaking by 2030 and achieving carbon neutrality by 2060. This goal is not only China's national policy in response to global climate change, but also a science-based national strategy, reflecting China's consideration of the actual situation and far-reaching planning for future development. Through this goal, China demonstrates its leadership and determination in global climate actions and its commitment to promoting the transition to green, low-carbon and sustainable development.

At present, although many countries have pledged to achieve carbon neutrality, public records show that no country has yet fully attained this goal. The Chinese government has proposed the ambitious goal of striving to achieve carbon neutrality by 2060, and has been actively and steadily advancing towards carbon peaking and carbon neutrality by building a carbon peaking and carbon neutrality "$1 + N$" policy system framework, promoting national renewable energy installation capacity and land greening, as well as other measures. Other countries such as Norway, Iceland, New Zealand and Costa Rica have also proposed their own carbon neutrality plans or have taken some measures to reduce carbon emissions, but there is also no public information showing that these countries have fully achieved carbon neutrality. Therefore, although many countries around the world are actively promoting carbon neutrality, no country has yet announced that it has achieved carbon neutrality.

In recent years, especially since the Russia-Ukraine conflict, Europe's reassessment of energy supply security has affected both the pace and policies of carbon emission reduction. This is exemplified by the United States' withdrawal from the Paris Agreement, Europe's restoration of some coal-fired power plants, and Mercedes-Benz's announcement in February 2024 that it would abandon its plan to fully electrify its vehicles by 2030. These measures, which are seen as emergency responses, are working against the long-term emission reduction goals, highlighting that geopolitical and economic factors may have a significant impact on climate policy in the global emission reduction process. Despite these challenges, governments and international organizations are still working hard to promote climate action to achieve the goals of the Paris Agreement. The realization of global net-zero emissions requires significant advancement in clean production technologies, major changes in clean energy, shifts in residents' behavior, and transition to an energy supply structure dominated by renewable energy. At the same time, governments must develop long-term policy frameworks that allow all government departments and stakeholders to plan for the reforms and promote an orderly transition.

2.1 European Union

The EU is committed to achieving the goal of carbon neutrality. Through key documents such as the European Green Deal and the European Climate Law, it has developed a comprehensive climate action plan. Table 3 lists the names and targets of the sectors that support the realization of the EU's carbon neutrality goal in the European Green Deal. These plans aim to drive the green transformation of various economic sectors, ensuring net-zero greenhouse gas emissions by 2050. To this end, the EU has taken a series of comprehensive measures, including the promotion and innovation of clean energy, the establishment of a circular economy, the green transformation of the construction industry, the creation of a sustainable transportation system, the establishment of a healthy and environmental friendly food system, the protection and restoration of the ecosystem, and the creation of a zero-pollution environment. These measures constitute the EU's roadmap to carbon neutrality. Through these efforts, the EU aims to steer the global economy to develop in a low-carbon, circular and sustainable direction, providing solutions to address climate change.

Table 3 The names and targets of the sectors supporting the achievement of carbon neutrality goals in the EU' s European Green Deal

ID	Sector	Target
1	Transportation	Reduce transport emissions by 29% by the third carbon budget compared to 2013, and by at least 70% by 2050. Build 100,000 electric vehicle charging stations and 100 hydrogen refueling stations by 2023
2	Residents' tertiary industry	Reduce carbon emissions by 54% by the third carbon budget compared to 2013, and by at least 87% by 2050
3	Agriculture	Reduce agricultural emissions by 12% by the third carbon budget, compared to 2013
4	Forest–Wood biomass	Offset 15–20% of national emissions
5	Industry	Reduce industrial emissions by 24% by the third carbon budget, compared to 2013, and by 75% by 2050
6	Energy supply	Renewable energy installed capacity should reach 20.1 GW by 2023 and 44 GW by 2028; by 2050, carbon emissions is expected to drop by 80–95% compared to 1990 levels
7	Waste management	Reduce transport emissions by 33% compared to 2013 by the third carbon budget and by at least 80% by 2050

2.2 *United Kingdom*

In order to ensure the realization of the carbon neutrality goal, the British government implemented multiple strategic plans and policies. Table 4 outlines key sectors and their targets supporting the realization of the carbon neutrality goal of the UK. In the field of energy, the British government has invested £240 million to the Net Zero Hydrogen Fund and £385 million to the Advanced Nuclear Energy Fund to support the development of clean energy. In the field of transportation, the UK plans to ban the sale of new petrol and diesel vehicles by 2030, and require a certain proportion of new car sales to meet ultra-low emission standards, with the goal of achieving zero emissions for all new cars by 2035. At the same time, the government will invest £4.2 billion in upgrading urban public transportation, £5 billion in bus, bicycle and pedestrian infrastructure, and £15 million in sustainable aviation fuel. In the construction sector, the "Future Home Standard" is implemented to improve energy efficiency. In addition, the UK has invested £1 billion to the CCUS (Carbon Capture, Utilization and Storage) Infrastructure Fund to support the research and development of zero- and negative-carbon technologies. These measures together form the strategic layout for the UK to achieve a green industrial revolution and industrial decarbonization.

2.3 *United States*

To achieve the goal of carbon neutrality, the United States has formulated strategic documents such as the *Plan for a Clean Energy Revolution and Environmental Justice* and the *Plan to Build a Modern, Sustainable Infrastructure and an Equitable Clean Energy Future.* Table 5 lists the names and targets of the sectors that support the realization of the carbon neutrality goal of the United States. The emission reduction strategy of the United States covers six key sectors: industry, construction, agriculture, residential consumption, energy, and transportation. Each sector has set specific emission reduction targets. These plans aim to comprehensively reduce greenhouse gas emissions through measures such as promoting the use of clean energy, improving energy efficiency, developing low-carbon technologies, and encouraging sustainable lifestyles, ultimately supporting the long-term goal of carbon neutrality.

3 Pathways to Achieving Park-Level Carbon Neutrality

With the advancement of China's "30–60" "Dual Carbon Goals", namely the carbon peaking and carbon neutrality strategies, the construction of zero-carbon parks has become the focus of many park developers, large state-owned enterprises and local governments. These parks are committed to using clean energy and minimizing

Table 4 The names and targets of the sectors supporting the UK' s carbon neutrality goals

Sector name	Prospects
Transportation sector	Developing offshore wind power with the target of achieving 40GW of offshore wind power installed capacity by 2030 Between 2023 and 2032, cut emissions by 21 million tonnes of carbon dioxide equivalent (CO_2e), equivalent to 5% of UK emissions in 2018 Offshore wind power networks could save consumers up to £6 billion by 2050, significantly reducing the environmental and social impacts on coastal communities
	Promote the development of low-carbon hydrogen and achieve a low-carbon hydrogen production capacity of 5 million kilowatts by 2030 Between 2023 and 2032, achieve a reduction of 41 million tCO_2e, equivalent to 9% of UK emissions in 2018 Reduce carbon emissions from heating and cooking through hydrogen blending fuels, without changing the domestic consumer experience, and reduce emissions from the gas used by 7%
	Accelerate the transition to zero-emission vehicles by providing subsidies to consumers who purchase electric vehicles, installing electric vehicle charging stations, and develop and mass-produce electric vehicle batteries Saving about 5 million tCO_2e by 2032 and 300 million tCO_2e by 2050
	Developing net-zero emission aircraft, sustainable aviation fuels (SAF) and clean ocean technologies Clean marine technologies could save up to 1 million tCO_2e by 2032, and sustainable aviation fuels could save nearly 15 million tCO_2e by 2050
Transportation sector	Between 2023 and 2032, green bus, bicycling, and walking could save about 2 million tCO_2e
Construction sector	Between 2023 and 2032, a reduction of 71 million tCO_2e will be achieved, equivalent to 16% of UK emissions in 2018
	600,000 heat pumps will be installed per year by 2028
	Houses built according to future housing standards will be "zero-carbon houses", with carbon dioxide emissions 75% to 80% lower than houses built according to current standards
Other	By 2030, the deployment of CCUS in four industrial clusters will be promoted to capture and store 10 million tCO_2e per year 40 million tCO_2e will be reduced between 2023 and 2032, equivalent to 9% of UK emissions in 2018

carbon emissions, and even achieving net-zero carbon emissions through various innovative practices. A zero-carbon park, or a carbon neutral park, refers to a park that achieves a balance between carbon emissions and absorption within a certain area through various measures. Although there is currently no unified standard for zero-carbon parks internationally, many places and organizations in China have begun to explore and implement the construction of low-carbon or zero-carbon parks, such as the first domestic renewable energy "carbon neutral" smart park certification obtained by Goldwind Science and Technology Yizhuang Smart Park, and other practical cases such as the Hisense Jiangmen Zero-Carbon Smart Park, the Ordos Zero-Carbon

Table 5 The names and targets of the sectors supporting the realization of carbon neutrality goals in the United States

Sector name	Prospects
Industrial sector	Let the American auto industry win in the twenty-first century with its own technology
	Achieve zero carbon emissions in the power sector by 2035
Construction sector	Investing heavily in improving building energy efficiency, including retrofitting 4 million buildings and building 1.5 million new affordable housing units
	Building modern infrastructure
Agricultural sector	Promoting sustainable agriculture and ecological protection
Residential Life sector	Ensuring environmental justice and equitable economic opportunity
Energy sector	Making historic investment in clean energy innovation
Transportation sector	Implementing the American Clean Cars Program

Industrial Park, the Qingdao Sino-German Ecological Park, and the Chongqing AI City Park.

In order to guide and regulate the construction of low-carbon parks, some local governments have issued relevant policies and standards. For example, the "Shanghai Low-Carbon Demonstration Creation Work Plan" issued by the Shanghai Ecological Environment Bureau provides suggestions on carbon emission accounting methods for low-carbon development practice areas and low-carbon communities, covering key points such as the scope, elements, methods and data sources for the accounting. The "Implementation Plan for the Pilot Construction of Near-Zero Carbon Emission Zones in Shenzhen" issued by the Shenzhen Ecological Environment Bureau and the Development and Reform Commission offers specific guidance for the pilot construction of near-zero carbon emission zones in terms of definition, application requirements, construction approach and carbon emission accounting methods. The introduction of these policies and standards provides an important reference for the planning, construction and evaluation of zero-carbon parks, which will help promote the optimization of energy structure, improvement of energy efficiency, use of renewable energy and reduction of carbon emissions in the parks, thus ultimately achieving the goal of carbon neutrality. With the introduction of more local policies and the accumulation of practical experience, the definition and standards of zero-carbon parks will gradually become clear, contributing Chinese wisdom and solutions to global low-carbon development.

3.1 Definition of Park Carbon Neutrality

The zero-carbon industrial park is a new type of park development model that fully integrates the concept of carbon neutrality into the planning, construction and operation of the park. This type of park relies on a zero-carbon operating system, sets carbon neutrality goals through precise carbon emission accounting, and plans specific pathways to achieve these goals. The park uses ubiquitous sensing technology to comprehensively monitor the carbon generation and reduction processes, and integrates energy conservation, emission reduction, carbon sequestration and carbon sink measures through digital means to achieve intelligent management. In the zero-carbon park, industries achieve low-carbon development, energy achieves green transformation, facilities are clustered for shared use, and resources achieve circular utilization. The goal is to achieve a self-balancing of carbon emissions and absorption within the park, deeply integrating production, ecology and life to form a sustainable model of new industrial parks.

At present, the concept of zero-carbon industrial parks has not been clearly defined in China, but with the advancement of the "Dual Carbon Goals", the definition and practice of zero-carbon parks are gradually becoming clear. The construction of zero-carbon parks focuses on the integration of industry and urban development empowered by zero-carbon, clean energy supply, energy efficiency improvement, coordinated carbon reduction through scientific and technological ecology, and the use of digital and intelligent technologies to promote smart management. At the same time, the investment strategy provides multi-dimensional guarantees and achieves green leadership through human-park interaction. The definition of a zero-carbon park emphasizes that all carbon emissions including Scope 3 during the operation of the park are neutralized to achieve net-zero emissions and provide users with zero-carbon products and services. It is worth noting that in the concept of the park's full life cycle, management of carbon emissions in the design and construction stage and the operation stage is separated. Since it is difficult to achieve zero carbon emissions in the construction stage in the short term, and considering the large number of existing parks, the zero-carbon parks that we usually refer to focus on carbon emission management in the operation stage to achieve long-term carbon neutrality goals.

3.2 Sources of Carbon Emissions in Parks

Carbon emissions in parks are categorized into three scopes: Scope 1 (direct emissions), Scope 2 (indirect energy-related emissions), and Scope 3 (other indirect emissions).

Scope 1 emissions are the greenhouse gases emitted directly into the atmosphere from the sources within the physical boundaries or assets directly controlled by the park, such as emissions from coal-fired boilers inside the park, gasoline vehicles owned by the park, etc. These emission sources are part of the park's direct operational activities, making them relatively easy to be identified and managed.

Scope 2 emissions include indirect emissions from purchased electricity and heat, that is, indirect emissions from the park's use of externally supplied electricity and heat. Although this part of the emissions does not come directly from the park, the park still bears corresponding carbon responsibilities when using these energy sources.

Scope 3 emissions are other indirect emissions, including all other emissions generated by the park's production and operational activities, such as property operations, employee commuting, upstream and downstream products (such as purchased equipment, office decoration, office supplies, etc.) and carbon emissions from the products of front-end suppliers. Scope 3 emission management is more complex because it involves the entire supply chain and value chain of the park's operations.

Achieving a zero-carbon park is not easy, as it requires comprehensive management of the park's direct and indirect emissions, including the control of carbon emissions from the products of front-end suppliers. However, some large parks can offset part of their carbon emissions by implementing carbon sink projects, such as afforestation and wetland protection, thus moving towards the zero-carbon goal. In addition, the park can further reduce carbon emissions and achieve carbon neutrality by adopting renewable energy, improving energy efficiency, and using low-carbon technologies and products.

3.3 Pathways to Achieving Park-Level Carbon Neutrality

The pathway that a park takes to achieve carbon neutrality is a process that is both comprehensive and systematic. Especially for newly built parks, the most effective approach is to start from the planning and design stage, so that the zero-carbon concept can be integrated into every corner of the park from the beginning. Generally speaking, achieving a zero-carbon park involves the following key aspects.

In terms of energy, a series of comprehensive measures are taken to ensure that the park achieves a high proportion of renewable energy use in its production and operation process. These include installing rooftop photovoltaic systems and photovoltaic carports in the park to generate electricity directly from solar energy; deploying small wind power generation facilities in suitable areas to capture wind energy and convert it into electricity; and purchasing certified renewable energy electricity to ensure that the electricity consumed by the park comes entirely from clean energy. In addition, parks with necessary conditions can build biogas cogeneration facilities and heat pump systems to achieve efficient recycling of energy. They can also be equipped with energy storage power stations and heat/ cold storage devices to balance energy supply and demand, ensuring the continuity and stability of the park's energy supply. Through these strategies, the park can not only reduce its dependence on fossil fuels and reduce its carbon footprint, but also improve its energy self-sufficiency, enhancing both the environmental and economic benefits.

In terms of construction, the strategy to achieve carbon neutrality in the park focuses on significantly reducing energy consumption and improving energy efficiency. This includes the use of advanced energy-saving technologies and materials in the design and construction of buildings, such as using energy-saving insulation materials to reduce heat loss, installing sunshades to reduce additional heat from sunlight, and using triple-glazed windows to improve thermal and acoustic insulation. All new buildings will be designed in accordance with green building standards to obtain the US Leadership in Energy and Environmental Design (LEED) Platinum certification, which represents a high standard of energy conservation, environmental protection and sustainability. Buildings in the park are equipped with smart meters that are connected to an intelligent energy management system for centralized monitoring of energy use. Such a system can optimize energy distribution, improve energy efficiency, and facilitate the management and maintenance. As a result of these measures, the buildings in the park can not only reduce their reliance on energy and reduce operating costs, but also provide a more comfortable and healthier environment for residents and workers, while minimzing the impact on the environment, helping to achieve the overall carbon neutrality goal of the park.

In terms of transportation, the key to achieving carbon neutrality in the park lies in promoting a comprehensive electrification transformation. This includes the extensive use of electric vehicles within the park and, to this end, the provision of adequate charging station facilities to meet the energy needs of electric vehicles. In addition, the establishment of a shared electric vehicle rental center can encourage more people to use electric vehicles, thereby further reducing the use of fossil fuels. The electricity used by the charging station will come from renewable energy sources such as wind power and solar power to ensure that the charging process is clean and sustainable. Using retired car batteries as energy storage devices can not only improve energy efficiency, but also promote battery recycling. The intelligent charging system can automatically adjust the charging period and power according to the load on the power grid and the energy supply, optimizing the use of electricity. The park can also introduce a variety of electric transportation tools such as driverless cars, electric sightseeing cars and shared bicycles to provide convenient and environmentally friendly travel options for people in the park. These measures together create a low-carbon, efficient and intelligent transportation system that will help to significantly reduce the park's transportation carbon emissions and promote the realization of carbon neutrality goals.

In terms of carbon sinks, the park's strategy for achieving carbon neutrality is to increase and create carbon sink projects in a variety of ways, thereby improving the park's carbon absorption capacity. This includes large-scale afforestation inside and outside the park and the planning and implementing greening projects, to offset the park's carbon emissions in a natural way. In addition, algae bioreactors can be installed on the exterior walls of the park's buildings to take advantage of the highly efficient carbon dioxide absorption capacity of algae during photosynthesis. Algae can not only effectively absorb carbon dioxide and reduce greenhouse gases in the atmosphere, but also absorb hazardous waste gases such as nitrogen dioxide. These algae can also be processed into green powders and used as nutritional additives in the cosmetics and food industries, providing both environmental and economic benefits. In addition to natural and biotechnological means, the park can also achieve carbon neutrality by participating in the carbon market and purchasing negative carbon products. This approach allows the park to offset its own carbon emissions through carbon credits or carbon offsetting projects, thereby achieving a net-zero increase in carbon emissions within the carbon emissions trading system. By adopting a combination of these methods, the park can increase carbon sinks at different levels, effectively offset carbon emissions through a combination of natural ecology, technological innovation and market mechanisms, contributing to achieving the goal of carbon neutrality.

In terms of management, the key to achieving carbon neutrality in the park lies in the comprehensive adoption of digital and refined management, using advanced information technology and intelligent systems to optimize resource allocation and operational efficiency. In terms of energy management, the park deploys building operation systems (EBO), power management systems (PEM) and intelligent micro-grid systems (EMA) to achieve effective synergy between energy supply (including wind power, geothermal, biogas and photovoltaic power, etc.), energy storage (such as large-capacity batteries, electric vehicle energy storage and thermal storage equipment) and energy demand (heat, cooling and electric load). This system integration improves the overall operational energy efficiency of the park and ensures carbon neutrality in the operation stage. In terms of operation and property management, the park achieves refined operation management through digital and intelligent means. This not only reduces the operation activities with energy consumption that results in no additional value, lowers energy waste, but also reduces the dependence on manpower. Through intelligent systems, the park can more effectively monitor and adjust energy use, optimize maintenance work, improve response speed and reduce management costs. In addition, the digital management of the park also includes waste management and water resource management. Through intelligent monitoring and data analysis, the efficiency of resource recovery and recycling can be improved and the waste of water resources can be reduced. This comprehensive digital transformation not only improves the sustainability of the park, but also brings economic savings and operational convenience to the park. Through these comprehensive management measures, the park can achieve efficient resource utilization and minimize environmental impact at all levels, pushing the park towards the goal of carbon neutrality.

Through the implementation of comprehensive measures, the park can indeed significantly reduce carbon emissions during its operation and achieve the goal of carbon neutrality. This process involves the optimization of energy structure, the improvement of the energy efficiency of buildings, the electrification of transportation, the increase of carbon sinks, and the digitization and intelligence of management. These measures not only have a positive impact on the environment, combating climate change by reducing greenhouse gas emissions, but also enhance the sustainability of the park and improve its market competitiveness. The park that implements these measures can also attract companies and investors who prioritize sustainable development and environmental protection to reduce operating costs and achieve economic benefits. In short, by achieving carbon neutrality, the park not only fulfils its social responsibility and protects the environment, but also lays a solid foundation for its own sustainable development.

4 Demonstration Cases of Park-Level Carbon Neutrality

4.1 Zhongqing New Energy Green Near-Zero Carbon Industrial Park

The construction of Zhongqing New Energy Green Near-Zero Carbon Industrial Park in Qingyun County, Dezhou City, Shandong Province was advanced from six dimensions: green power supply, green buildings, green transportation, a smart energy carbon management platform, diversified carbon sinks and zero carbon certification. Relying on the rich solar and wind energy resources, the park introduces green power transactions from existing wind power projects, taps into the potential of new wind power and photovoltaic projects to create an "integrated source-grid-load-storage" supply model, thus achieving 100% green power coverage in the park.

Qingyun County has significant advantages in green power resources, and the total installed capacity of wind power and photovoltaic power ranks first in the city. In 2022, Qingyun County's new energy power generation reached 1.14 billion kWh, about 1.2 times the total electricity consumption of the whole society, which provides favorable conditions for the development of the energy storage industry. The construction of the Three Gorges Qingyun Energy Storage Power Station Demonstration Project and the upcoming Phase II project will further consolidate Qingyun County's position in the field of energy storage. Energy storage development has formed a highland in Qingyun County. With the commissioning of the Phase II demonstration project of the Three Gorges Energy Qingyun Energy Storage Power Station, the total capacity of the Qingyun Power Station will reach 301 MW/602 MWh (maximum charging and discharging power of 301 MW/total storage capacity of 602 MWh), providing strong support for the State Grid Shandong Company to create the "Qingyun Whole-region Green Power Supply Demonstration Zone". At the same time, Qingyun County also pays attention to the innovation in scientific research and talent training in energy storage, and cooperates with Shandong University of Science and Technology, Three Gorges Group and other organizations to jointly build an energy storage college, promoting the R&D and application of energy storage technologies. The sub-regional planning of the Zhongqing New Energy Green Near-Zero Carbon Industrial Park (Area A: battery cells and PACK area; Area B: energy storage system production area; Area C: battery material production and cascade recycling area; Area D: sodium-ion battery area) will help form a highly concentrated industrial cluster, promote upstream and downstream collaboration, and improve the overall efficiency of the industrial chain. The construction and development of the park will bring significant economic benefits to Qingyun County, with an estimated annual output value of CNY 11.2 billion. Overall, Qingyun County has embarked on a solid path to break the industrial dilemma with new energy by clarifying the direction of industrial development, planning and building a green near-zero carbon industrial park, taking advantages of green power resources, and promoting the development of energy storage. These measures not only help Qingyun County achieve its carbon neutrality goals, but also inject new vitality into local economic development.

The construction of the near-zero carbon smart park is a comprehensive and advanced planning process. It aims to build a zero-carbon energy system with a high proportion of green power supply, so as to shape the brand image of the park and attract outstanding enterprise, jointly building the architecture of the near-zero carbon industry. The park will support enterprises to obtain green supply chain certification for their products, creating zero-carbon products with international competitiveness, thereby consolidating the production capacity base of the park. In the short term, the park will ensure 100% green power supply by purchasing green power, carry out energy-saving renovation of existing buildings, initiate certifications such as green factories and Environmental Product Declarations (EPDs), and conduct carbon footprint assessments. At the same time, the park will also formulate medium- and long-term zero-carbon plans to lay the foundation for sustainable development. The medium- and long-term goals include developing new renewable energy projects, constructing green buildings, promoting green transportation, adopting smart energy and carbon management platforms, and developing diversified carbon sink projects to achieve the park's carbon neutrality goals. In addition, the park will pursue zero-carbon certification to further enhance its status and influence in the field of sustainable development. Through these comprehensive measures, the near-zero carbon smart park will become an important platform for promoting the development of a low-carbon economy and contribute to achieving regional and even global carbon neutrality goals.

a. Green power supply. Encourage the nine wind farms that have been built in Qingyun as power generating enterprises and the Zhongqing New Energy Green Near-Zero Carbon Industrial Park as an power user to participate in green power market transactions, and sign electricity spot transaction or medium- and long-term transaction contracts at power sales companies or power grid companies through bilateral negotiations. Make full use of the park's land resources and rooftop resources to build distributed wind power and rooftop distributed photovoltaic systems, achieving self-generation and self-use in the park. At the same time, use the available land around the park to develop centralized photovoltaic and wind power projects, introduce green power directly into the park, and reasonably allocate a certain proportion of energy storage facilities to create a "source-grid-load-storage integration" project. This approach can promote the greening of the park's energy use, while effectively reducing the park's electricity costs. Figure 1 and Fig. 2 are the cases of rooftop distributed photovoltaic power generation systems and distributed wind power operations in the park, respectively.

Fig. 1 Case of rooftop distributed photovoltaic power generation system in a park

Fig. 2 Case of distributed wind power operations case in a park

b. Green buildings. Optimize and upgrade existing buildings by carrying out energy-saving renovations on existing factories and buildings in the park; strictly manage incremental buildings by promoting integrated building design, load optimization, and ultra-high energy-efficiency intelligent equipment, thus reducing building loads and improving building equipment efficiency., and ultimately achieving near-zero carbon emissions from buildings in the park.
c. Green transportation. An integrated "photovoltaics-storage-charging-swapping" station is built in the park, as shown in Fig. 3. This station operates on the principle of prioritizing the consumption of photovoltaic-generated power, storing excess energy, feeding surplus energy into the grid after full storage, and implementing

Fig. 3 "Photovoltaics-storage-charging-swapping" integrated station

night-time charging and day-time discharge of energy storage, thus achieving clean energy storage and local consumption.

d. Smart Energy and Carbon Management Platform. As shown in Fig. 4, the Smart Energy and Carbon Management Platform can realize the timely, complete and standardized collection and storage of carbon management data; efficiently process the carbon management data of the park, enabling the integration and intelligence of carbon management data processing; dynamically meet the carbon management needs of various users, including compliance with policies and supervision, and refined management of carbon emissions and carbon assets, providing a basis for carbon trading.
e. Diversified carbon sinks. Taking the comprehensive improvement and transformation project of the park's ecological environment as a starting point, green corridors will be built in the park to increase vegetation coverage and enhance the carbon sequestration capacity of trees; the carbon sink capacity will be enhanced through comprehensive greening, three-dimensional greening of buildings, and the greening of roads.

4.2 Ordos Zero-Carbon Industrial Park

Ordos is the world's largest coal-producing, coal-fired power and coal-chemical city, representing a fossil energy-based industrial system. Under the background of "Dual Carbon Goals", the governments of the Inner Mongolia Autonomous Region

Fig. 4 Smart energy and carbon management platform

and Ordos City have turned challenges into opportunities. They make full use of the rich local renewable energy resources and complete industrial manufacturing system, taking the lead in building a next-generation power system in the country. This initiative aims to gradually realize the decarbonization of the industrial system, ushering in a new era of zero-carbon industrial parks. They are committed to becoming a best-practice model for China's traditional industrial areas to achieve zero-carbon transformation, and contributing Chinese solutions to the zero-carbon transformation of the global industrial system.

The Ordos Zero Carbon Industrial Park Demonstration Project consists of a zero-carbon power supply system and a smart power consumption park. The zero-carbon power supply system consists of wind power, photovoltaic power generation, and energy storage systems. Substations and collection lines are built to supply power to the Ordos Zero-Carbon Industrial Park. Supported by modern information technology, big data, energy storage and other technologies, the industrial park is equipped with supporting energy storage power stations and plans to have a supporting electrolyzers for green hydrogen production. It mobilizes the load-side regulation and response capabilities of the park to establish a multi-directional, interactive and highly integrated development route for source, grid, load and storage. The project uses a source-grid-load-storage coordination control system to achieve collaborative management of the source, grid, load and storage, thereby better matching output and load. The project tracks carbon footprints through the "Ark" carbon management system, and gradually creates a domestic "zero-carbon industrial park" model using methods such as energy management, self-use of green power, substitution carbon with hydrogen, and carbon trading. In January 2024, the Ministry of Industry and Information Technology announced the list of the first batch of typical application scenarios and cases of industrial green microgrids in China. Ordos

Zero Carbon Industrial Park (Mengsu Zero Carbon Industrial Park) was successfully selected into the list, becoming one of the first three industrial parks with typical applications of industrial green microgrids in China.

Ordos Zero Carbon Industrial Park is located in the western area of Jiangsu Industrial Park within the Mengsu Economic Development Zone of Ejin Horo Banner. It is the world's first 100 billion-level zero-carbon industrial park built by the Ordos Municipal Government, Ejin Horo Banner Government and Envision Group. The planning of the Ordos Zero Carbon Industrial Park: The overall layout of the Zero-Carbon Industrial Park is based on "one axis, two cores and two zones", with zero-carbon high-tech industries as the display axis, supported by commercial business core, scientific and technological innovation core, industrial zone and living service zone. It focuses on realizing the zero-carbon concept by building an innovative demonstration system of "1 + 2 + 5 + 8" (a new power system, a set of international standards and an energy carbon management platform, five zero-carbon industrial chain clusters, and eight innovative demonstration new energy systems).

Ordos Zero-Carbon Industrial Park specifically realizes the zero-carbon concept by building a "1 + 2 + 5 + 8" innovation demonstration system.

"1" is to build a new power system. Relying on the advantages of the wind and solar resources in the Ordos Plateau, wind and solar power plants are planned and built within a 150-km radius of the Zero Carbon Industrial Park to ensure a continuous supply of green electricity for the Zero Carbon Industrial Park. By building a microgrid, 80% of the green power can be directly supplied with the green electricity generated inside the park itself and 20% of the green power can be traded on the grid.

"2" means publishing a set of international standards and building an energy and carbon management platform. Envision has joined hands with influential international and domestic authoritative organizations such as Bureau Veritas of France and China National Institute of Standardization to jointly improve the international standards for zero-carbon industrial parks. At present, the three local standards of "Green Power Application Evaluation Method", "Construction Specification for Zero-Carbon Industrial Park" and "Specification for the Carbon Accounting of Zero-Carbon Industrial Parks System" have become the first batch of standards for provincial zero-carbon industrial parks issued nationwide. The first national standard related to zero-carbon industrial parks, "Guidelines for the Construction of Low-Carbon Industrial Parks", led by enterprises within the park, has been approved by the National Standardization Administration. 52 enterprises of designated size have been connected to the Internet of Things on the Ark pilot platform for operation, with plans to gradually promote it throughout the city, achieving intelligent analysis and dynamic supervision of energy and carbon emissions, and striving to promote the application of "zero-carbon green code" in 2023 to lay a solid foundation for international "zero-carbon trade".

"5" refers to the construction of a cluster of five zero-carbon industrial chains of "wind-solar-hydrogen-vehicle-storage". At present, an industrial development pattern has been formed with leading enterprises serving as models and upstream and downstream industries providing supporting services.

"8" means that by the end of the 14th Five-Year Plan period, the "wind-solar-hydrogen-vehicle-storage" cluster will reach an output value of CNY 100 billion. Ultimately, eight innovative demonstrations will be formed, including fully green power supply, high-proportion new integrated energy storage system, intelligent source-load interactive microgrid, intelligent IoT energy-carbon management platform, international zero-carbon industrial park standards, and the "wind-solar-hydrogen-vehicle-storage" zero-carbon industrial chain cluster, the "Silicon Valley" for the cultivation of green technology experts, and models for promoting zero-carbon industrial parks nationwide and globally.

The zero-carbon industrial park built by Envision Group in Ordos follows the positioning of "Chinese model and world benchmark", and aims to achieve "CNY 300 billion of green new industrial output value, 100,000 green high-tech jobs, and 100 million tonnes of annual carbon dioxide emission reduction" by the end of the 14th Five-Year Plan period. The Envision Ordos Zero-Carbon Industrial Park was selected in the COP27 "2022 Corporate Climate Action Case Collection" and it was reported by the World Economic Forum in the "Industrial Cluster's Transformation to Net Zero Emissions" report. At the same time, at COP28, with its significant contribution to global decarbonization and emission reductions, the Ordos Zero-Carbon Industrial Park stood out among more than 1,000 projects worldwide and won the Energy Transformation Transformer Award. The Ordos Zero-Carbon Industrial Park innovatively realizes the effective integration of green zero-carbon energy production and load, which not only promotes the construction of new power systems, but also facilitates the development of green industrial systems. It is currently expanding to Europe, the Middle East and Southeast Asia, contributing China's innovative solutions to the zero-carbon industrial transformation in the globe.

Chapter 4
Analysis of the Pathways to Achieving Organizational Carbon Neutrality

Carbon neutrality at the organizational level is the key to addressing global climate change. Organizations are social entities composed of individuals or groups with common goals and clear boundaries. In the context of carbon emissions, enterprises, as a form of organization, have become the main source of global greenhouse gas (GHG) emissions because their production activities consume a large amount of materials and energy. The sources of greenhouse gas emissions generated by human social and economic activities are mainly energy production and industrial manufacturing processes, which are mainly governed by energy and industrial enterprises. Therefore, organizations play a key role in greenhouse gas emissions and reductions. Greenhouse gas emissions at the organizational level are the main source of emissions. Organizations are the main entities responsible for reducing greenhouse gas emissions and achieving carbon neutrality. If major organizations in the world achieve carbon neutrality, global carbon neutrality can basically be achieved.

1 Characteristics of Organizational Carbon Neutrality

Organizations play a complex and diverse role in greenhouse gas emissions. They are not only the sources of emissions, but also key forces in driving the low-carbon transformation. Different organizations have different sources of greenhouse gas emissions due to their specific production methods, process technologies and the raw and auxiliary materials used.

Emissions from steel production enterprises come mainly from the use of coke for reduction reactions during blast furnace ironmaking. This process not only consumes a lot of energy, but also generates significant CO_2 emissions.

Emissions from cement production companies come mainly from the combustion of coal during the clinker calcination process and the CO_2 naturally generated by the decomposition of limestone at high temperatures.

C. Chen et al., *Sustainable Carbon Neutrality*,
https://doi.org/10.1007/978-981-95-1691-9_4

Emissions from electronics assembly companies are mostly indirect emissions, mainly from the electricity used in the production process, which may come from the combustion of fossil fuels.

Emissions from power generation companies come directly from the process of burning fossil fuels to generate electricity and are one of the main sources of carbon emissions in the energy industry.

The diversity in these emission sources requires organizations to take targeted measures to reduce emissions, such as improving processes, enhancing energy efficiency, using clean energy, and adopting carbon capture, utilization, and storage technologies. At the same time, organizations should also actively invest in research and development to drive technological innovation, and optimize supply chain management, thus reducing the carbon footprint of the entire production cycle. Achieving carbon neutrality at the organizational level requires a comprehensive consideration of both direct and indirect emissions. The goal of net-zero emissions can be achieved by formulating scientific carbon reduction strategies and implementing effective carbon offsetting measures, such as increasing carbon sinks through afforestation or purchasing carbon credits. Through these efforts, organizations can not only mitigate their own environmental impact, but also contribute to the common goal of achieving carbon neutrality on a global scale.

Organizations play a vital role in achieving the global carbon neutrality goal by reducing greenhouse gas emissions and promoting the development of a low-carbon economy through a series of specific actions. These actions cover the following four aspects.

(1) Research and development of energy-saving technologies: Organizations should be committed to developing and deploying advanced energy-saving technologies that can improve the efficiency of energy use and reduce energy waste, thereby reducing greenhouse gas emissions.
(2) Produce efficient products: By designing and manufacturing products that are both energy efficient and environmentally friendly, organizations can ensure that products are more energy efficient during their usage, leading to reduced energy consumption, and consequently, less greenhouse gas emissions.
(3) Invest in clean energy: By investing in clean energy projects such as solar and wind power, organizations can not only reduce the use of fossil fuels in their own operations, but also promote the overall development of clean energy and provide society with cleaner energy options.
(4) Optimize supply chain management: By implementing green supply chain management, organizations make sure carbon footprint reduction is considered at every stage, from raw material procurement to product distribution, thereby promoting the sustainability of the entire supply chain.

Through these measures, organizations can not only improve their own environmental performance, but also drive the upstream and downstream of the supply chains and the entire industry as a whole to develop in a low-carbon and environmentally friendly direction through demonstration effects and market influence. This top-down change will help promote the transformation of the entire society as a whole to a low-carbon economy and achieve the global goal of carbon neutrality. If major global organizations can achieve carbon neutrality, that is, achieve net-zero emissions through emission reduction and carbon offsetting measures, the goal of global carbon neutrality will be more likely to be achieved. This requires not only the active participation of organizations, but also the synergy of policy support, market incentives and technological innovation.

2 Pathways to Achieving Organizational Carbon Neutrality

The pathways to achieve carbon neutrality at the organizational level includes four main steps, namely, organizational commitment, quantification of carbon emissions, voluntary reduction of emissions, and offsetting of carbon emissions. These four steps provide the basic framework for organizations to achieve carbon neutrality. Organizations need to continuously monitor and evaluate their progress towards carbon neutrality and adjust strategies and measures according to the actual condition to ensure that the goal of carbon neutrality is ultimately achieved. At the same time, organizations should also enhance stakeholder engagement, improve transparency, and demonstrate their sense of responsibility and leadership in addressing climate change.

2.1 Organizational Carbon Neutrality Commitment

An organization's commitment to carbon neutrality is an important step in demonstrating its commitment to environmental responsibility and sustainable development. This commitment not only demonstrates the organization's determination to reduce greenhouse gas emissions, but also reflects its clear goal of achieving net-zero emissions. A complete organizational commitment should include specific plans for achieving its carbon neutrality goals, including setting clear timelines, developing action plans to achieve these goals, and possible emission reduction measures and strategies. In addition, this commitment should also include the transparency in the quantification, monitoring, reporting and verification (MRV) processes of carbon emissions to ensure that the organization's efforts and progress can be monitored and recognized by internal and external stakeholders. Through this comprehensive and transparent commitment, an organization can build trust and motivate its members, partners and all sectors of society to work together, contributing to the achievement of global carbon neutrality goals. A complete organizational commitment should include the following:

- The boundaries of carbon neutrality, which refer to the physical boundaries and scope of greenhouse gas accounting under the carbon neutrality goal.

Table 1 Ant Group's carbon neutrality goals and action plans

• Achieve carbon neutrality for operational emissions (Scope 1 and 2) in 2021 • Achieving net-zero emissions by 2030 • Regular disclosure of carbon neutrality progress starting 2021
Ant Group's Carbon Neutrality Action Plan • Promote the construction of green parks, carry out energy-saving and emission reduction renovations in the existing parks to improve energy efficiency; design, build and operate new parks in accordance with green building standards • Encourage employees to work and travel in a green way • Actively and steadily promote green investment and jointly build a "Carbon Neutrality Technology Innovation Fund" • Promote emission reduction in the supply chain, actively adopt new technologies such as liquid cooling, promote emission reduction in data center, establish a green procurement mechanism, and comprehensively promote paperless procurement • Strengthen the scientific management of greenhouse gas emission control, use Ant Chain to store the evidence of carbon emission accounting and carbon neutrality processes, and disclose carbon neutrality results regularly • Carefully evaluate and use carbon offsetting options, invest in forests and other nature-based solutions, and plant "carbon sink forests" in the name of employees starting from 2021

- Types of greenhouse gases, specifying whether the accounting covers CO_2 only, or covers all types of greenhouse gases.
- Timetable for achieving carbon neutrality.
- Means of emission reduction, which refers to the technologies and measures planned to achieve greenhouse gas emission reductions.
- The carbon offsetting strategy employed, including an estimate of the amount of carbon emissions to be offset, the nature of the offsetting, and the type of carbon credits.

The commitment to carbon neutrality should be made public through the organization's official website or social media. If an organization intends to participate in carbon neutrality programs such as the "Climate Neutral Now" initiative of the United Nations Framework Convention on Climate Change or the Science Based Targets initiative (SBTi), it can fill out a commitment letter as required and submit it to the Carbon Neutrality Program Management Committee after approval by the organization's top management. Taking Ant Group as an example, Table 1 lists the company's carbon neutrality goals and action plans.

2.2 *Quantification of Organizational Emissions*

Quantifying organizational emissions is a key step in achieving carbon neutrality, which requires organizations to form a complete and accurate greenhouse gas emissions inventory. This process involves a detailed accounting of all greenhouse gas emissions generated by the organization, including direct emissions (Scope 1), indirect emissions from energy procurement (Scope 2), and other indirect emissions (Scope 3), such as emissions in the supply chain.

Quantifying greenhouse gas emissions enables organizations to clearly understand their own emission levels, thereby identifying potential areas and priorities for emission reduction. This step is the basis for developing effective emission reduction strategies and a prerequisite for monitoring and reporting on emission reduction progress. When quantifying emissions, organizations usually follow internationally recognized standards and guidelines, such as ISO 14064-1:2018 Greenhouse gases—Part 1: Specification with guidance at the organization level for quantification and reporting of greenhouse gas emissions and removals, which is currently the most commonly used international standard for quantifying and reporting organizational greenhouse gas emissions. In addition, the Greenhouse Gas Protocol Corporate Accounting and Reporting Standards jointly issued by the World Business Council for Sustainable Development (WBCSD) and the World Resources Institute (WRI), also provide a widely recognized accounting framework. By following these standards and guidelines, organizations can ensure that the quantification process of their emissions is transparent, consistent and reliable. This not only helps the organization's internal management and decision-making, but also increase the trust of external stakeholders in the organization's emission reduction efforts. When the emission has been quantified, the organization can develop practical measures to reduce emissions and take actions to achieve carbon neutrality goals.

Greenhouse gas emissions quantification is usually carried out with the following steps.

a. Determining the Accounting Boundary

Organizations have different legal and organizational structures for their business activities. Firstly, an organization should define the boundaries of its GHG accounting by using the selected methodology to define the scope of the organization's business activities and operations in order to account for and report on the organization's GHG emissions. There are generally two methods for defining the boundaries of greenhouse gas accounting: the equity ratio method and the control method.

(i) Equity ratio method: An organization accounts for GHG emissions based on its equity ratio in a business. For example, if an organization owns 50% of a business, then 50% of the GHG emissions generated by the business are included in the organization's accounting scope.
(ii) Control approach: An organization accounts for all greenhouse gas emissions within the business scope of its control, and does not account for greenhouse gas emissions from the businesses in which it has an interest but does not hold control. Control may be defined from a financial or operational perspective. When using the control approach to consolidate greenhouse gas emissions, the organization must choose between the two criteria of operational control or financial control.

(a) Financial control: If a company has financial control over its business, it can directly influence its financial and operating policies and obtain economic benefits from its activities.
(b) Operational control: If a company or its subsidiaries have full power to propose and implement the operating policies of a business, then the company has operational control over the business.

The operational control method is commonly used to determine the accounting boundary of an organization's greenhouse gas emissions, that is, the organization accounts for the greenhouse gas emissions generated by the business over which it has operational control, and excludes the business over which it does not have operational control from the accounting boundary.

b. Identifying Greenhouse Gas Emission Sources

Greenhouse gas emissions generally come from the following sources.

Stationary combustion: The combustion of fuels in stationary facilities such as boilers, furnaces, burners, turbines, heaters, incinerators, engines and combustion towers.

Mobile combustion: The combustion of fuels in transportation vehicles such as cars, trucks, buses, trains, airplanes, steamships, ships, barges, etc.

Process emissions: Emissions from physical or chemical processes, such as CO_2 from clinker calcination in cement production, CO_2 from catalytic cracking in petrochemical processes, and perfluorocarbons from aluminum smelting.

Fugitive emissions: Intentional and unintentional leaks from equipment joints, seals, packaging and gaskets, as well as fugitive emissions from air conditioning refrigerants, wastewater treatment, maintenance areas, cooling towers, various gas treatment facilities, etc.

The greenhouse gases currently recognized by the United Nations Framework Convention on Climate Change include CO_2, CH_4, N_2O, HFCs, PFCs, SF_6 and NF_3. Organizations' greenhouse gas quantification should typically include the above seven greenhouse gases and convert them into tonnes of carbon dioxide equivalent (tCO_2e) according to the global warming potential (GWP) of each type of gas.

According to the latest ISO 14064-1:2018 standard, an organization's greenhouse gas emissions can be divided into the following six categories.

Category 1: Direct greenhouse gas emissions, including combustion emissions from stationary sources, combustion emissions from mobile sources, emissions from industrial process, fugitive emissions, etc. within the organizational boundary.

Category 2: Indirect emissions from imported energy, i.e. indirect emissions from the input of electricity, heat, etc.

Category 3: Indirect emissions from transportation, including emissions from the transportation of raw and auxiliary materials from suppliers to the organization, emissions from the transportation of the organization's products to downstream customers, and emissions generated by employee commuting and business travels.

Category 4: Indirect emissions from products and services used by the organization, including emissions from the manufacturing process of products purchased by the organization, emissions from the treatment of solid or liquid waste output by the organization, etc.

Category 5: Emissions associated with the use of the organization's products, including emissions related to the use of products sold by the organization throughout their life cycle, emissions related to the scrapping process of products sold by the organization, etc.

Category 6: Other indirect emission sources, special emission sources that are not included in the above five categories. This type of emission is defined by the organization itself.

In the first edition of the ISO 14064-1 standard (published in 2006), Category 1 mentioned above is defined as "Scope 1" direct emissions, Category 2 is defined as "Scope 2" indirect emissions, and Categories 3, 4, 5 and 6 are defined as "Scope 3" other indirect emissions. Scope 1 and Scope 2 emissions are mandatory reporting items, while Scope 3 emissions are optional reporting items.

The latest version of the standard (the second edition, released in 2018) abandons the concept of scope and divides organization-related emissions into six categories, reflecting the focus on emissions from the upstream and downstream of the value chain that organizations can influence. Organizations should account for all categories of emissions as accurately as possible when data is available.

Due to differences in operations, production processes, products and services, the proportion of emissions in the above categories varies greatly. When quantifying the greenhouse gas emissions of an organization, the proportion of emissions in each category should be estimated first. Based on these proportions, the organization can identify major emission sources, moderate emission sources and minor emission sources. Major and moderate emission sources should be accounted for in detail, and minor emission sources can be quantified as accurately as possible if data is available. Minor emission sources with unavailable data or poor data accuracy may be omitted according to appropriate principles.

Organizations may consider the following principles when selecting emission sources for quantification.

(i) 100% coverage of category 1 emissions (direct emissions) and category 2 emissions (indirect emissions from imported energy) relevant to the organization.
(ii) Other emission categories with estimated proportions exceeding 1% of the organization's total emissions should be taken into account unless such quantification is not technically feasible or cost-effective.
(iii) Emission sources with an estimated proportion less than 1% of the organization's total emissions may be excluded from the accounting, but the excluded emissions shall not exceed 5% of the organization's total emissions.

c. Selecting a Quantification Method

Greenhouse gas emissions are generally calculated using the emission factor method and the material balance method. The most common method for calculating greenhouse gas emissions is the use of emission factors. Emission factors can refer to the General Guideline of the Greenhouse Gas Emissions Accounting and Reporting of

Industrial Enterprises, the national industry-specific guidelines for greenhouse gas accounting methods and reporting, and the IPCC assessment reports.

For emissions from specific process flows, in the absence of industry-specific calculation tools or guidelines, the emissions can be calculated using the material balance method or stoichiometric method based on specific facilities or process flows.

Selection of Quantification Method

Fuel combustion is the most common source of greenhouse gas emissions, and its quantification methods can be simple or complex.

$$\text{GHG emissions} = \text{activity data} \times \text{emission factor}$$

Simple Version:

- Activity data comes from the settlement statements of fuel suppliers, and the annual data is the sum of the settlement data within the period.
- The emission factors are derived from IPCC default values and do not take into account unburned carbon or other greenhouse gas emissions (such as CH_4).

Complex Version:

- Activity data (such as natural gas consumption) is measured using two sets of parallel flow meters, which simultaneously measure temperature and pressure, and calculate the standard volume of gas through built-in electronic modules with a measurement error $< 1.5\%$.
- The gas composition is measured using a gas chromatograph, with 4–8 samples per hour according to ISO 10715.
- Hourly or daily emission factors are determined based on measured levels of methane and 10 other gases.
- The entire measurement system is automatically calibrated every day and regularly calibrated every month.
- Gas calibration meets the requirements of ISO/IEC 17025 standard, and gas chromatograph manufacturers must be certified for ISO 9001.
- The chromatographs are validated annually by a laboratory accredited to ISO/IEC 17025 in accordance with ISO 10723.

d. Calculation of Greenhouse Gas Emissions

The organization calculates greenhouse gas emissions for each source and category based on the selected quantification method, and then aggregates the calculated data at the organizational level in a standardized format so that the data collected by different business units and facilities are comparable. There are two basic methods for aggregating greenhouse gas emissions, and organizations can choose an appropriate aggregation method.

Centralized approach: Individual facilities report activity data/fuel use data (e.g. fuel consumption) to the organization, which then aggregates greenhouse gas emissions at the organizational level.

Decentralized approach: Facilities collect activity data/fuel use data and directly calculate their greenhouse gas emissions using a unified methodology, which are then reported at the organizational level.

e. Prepare Organizations' Greenhouse Gas Emission Inventories and Emission Reports

On the basis of the above steps, the organization's greenhouse gas emissions inventory is compiled. If allowed under the conditions, the organization may establish a quality management system for the greenhouse gas emissions inventory to manage the greenhouse gas inventory results and ensure that the greenhouse gas inventory results are accurate, transparent and traceable. The general measures taken by organization to manage the quality of the greenhouse gas inventory shall include the following aspects.

(1) Data collection, input and processing activities

- Check a sample of the input data for transcription errors.
- Identify necessary changes to electronic worksheets to make them more suitable for quality control or quality inspection.
- Ensure that appropriate version control is implemented on executed electronic documents.

(2) Data recording

- Make sure all raw data in the spreadsheets are indexed to their source.
- Check that copies of the references cited have been filed.
- Check that the assumptions and criteria used to select the boundary, base year, methodology, activity data, emission factors and other parameters are documented.
- Check that the changes in data or methodology have been documented.

(3) Calculate emissions and verify calculation process

- Check that emission units, parameters and conversion factors are appropriately labelled.
- Check that units are appropriately labelled and correctly applied throughout the calculation process.
- Check that the emission factors are correct.
- Check the data processing steps (such as formulas) in the spreadsheet.
- Check that a clear distinction is made between input data and calculated data in the worksheet.
- Manually or electronically check the calculation process on a representative sample.
- Check some of the calculations by simplifying them.
- Check data aggregation by source category, business unit, etc.
- Check inputs and calculations for consistency in time sequence.

Organizations prepare formal greenhouse gas emissions reports based on the results of greenhouse gas inventory, which typically include the following contents.

- Description of the organization.
- Personnel responsibilities.
- The period covered in the report.
- Description of organizational boundaries.
- Description of the reporting boundaries, including the criteria defined by the organization to define major sources of emissions.
- Direct GHG emissions, with results of separate quantification of CO_2, CH_4, N_2O, NF_3, SF_6, HFCs, PFCs, and the results of conversion to tonnes of carbon dioxide equivalent (tCO_2e).
- A description of how biogenic CO_2 emissions and removals are handled and quantified separately in the GHG inventories.
- Direct GHG removals (if quantified).
- Justification for exclusion of major GHG sources and sinks.
- Indirect GHG emissions for each category are quantified separately.
- The selected historical base year and the base year GHG inventory.
- Explanation of any changes to the base year or recalculation of the base year inventory, and any limitations on comparability resulting from the recalculation.
- Reference sources or descriptions of the quantitative methods and the rationale for their selection.
- A description of the change from the previously chosen quantification method.
- Reference sources or descriptions of GHG emission or removal factors.
- The impact of uncertainty in each GHG emission or removal category on the final results.
- Uncertainty assessment and its results.
- A statement that the GHG report conforms to this standard.
- Disclosure of whether the GHG inventory, report or statement has been verified, as well as the type of verification and the level of assurance achieved.
- The GWP values and their sources, if not from the IPCC assessment reports on climate change, should state the reference database of the emission factors and their source.

Greenhouse gas inventories and reports typically cover the previous calendar year, with the first complete greenhouse gas inventory accounting year being the base year, and subsequent annual inventory data being compared with the base year. The base year inventory should represent the normal production level of the enterprise. If there are major changes in the organizational structure, such as mergers and reorganizations, the base year inventory should be recalculated.

2.3 *Voluntary Emission Reduction Actions*

Once an organization has quantified its GHG emissions, voluntary emission reduction actions become the cornerstone in achieving the carbon neutrality goal. This requires organizations to take a series of practical measures to reduce greenhouse gas emissions based on a comprehensive understanding of their own emissions. These measures include but are not limited to the following aspects.

- Improve energy efficiency: Reduce energy consumption and improve energy efficiency by optimizing operating processes and upgrading technologies.
- Adopt clean energy: Switch to renewable energy such as solar, wind, geothermal, and so on, thus reducing the dependence on fossil fuels.
- Improve processes: Update production technologies, adopt more environmentally friendly production processes, and reduce emissions during the production process.
- Upgrade equipment: Invest in efficient, low-carbon equipment and replace old, high-emission equipment.
- Optimize supply chain: Reduce the carbon footprint in the supply chain by choosing environmentally friendly suppliers and logistics methods.

Under the overall goal of carbon neutrality, organizations need to formulate specific annual GHG reduction targets and management plans to ensure the continuity and effectiveness of emission reduction actions. This usually involves the establishment of a systematic carbon emission management system that includes monitoring, reporting and verification of carbon emissions within the organization, as well as evaluation and tracking of the effectiveness of emission reduction measures. Figure 1 shows a schematic diagram of voluntary emission reductions by corporates to ahieve carbon neutrality. Through these voluntary emission reduction actions, organizations can not only directly reduce their own greenhouse gas emissions, but also promote the low-carbon transformation of the entire industry and even society through demonstration effects and market influence. These bottom-up efforts are crucial to achieving the goal of carbon neutrality.

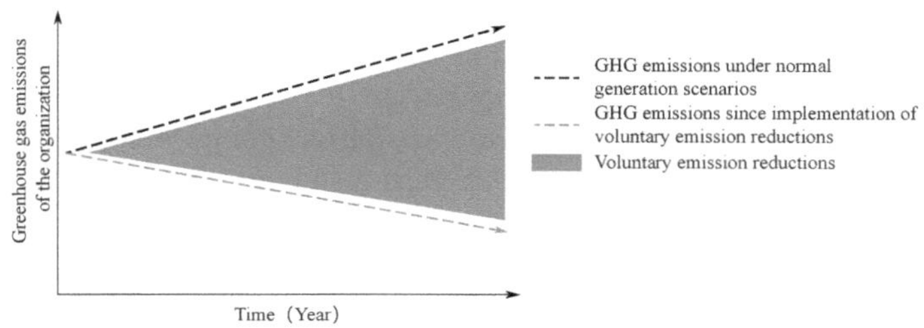

Fig. 1 Schematic diagram of voluntary emission reductions by corporates to achieve carbon neutrality

Based on the quantification results of the organization's greenhouse gas emissions, we can gain an in-depth understanding and diagnosis of the organization's energy consumption and greenhouse gas emissions, and formulate and implement a voluntary emission reduction plan to achieve carbon neutrality. Within the scope that is under the control of the organization, the voluntary emission reduction plan to achieve carbon neutrality can be roughly based on the following five directions.

(1) Establish and improve the energy management system and carbon emission management system: Install effective energy monitoring and metering equipment, and establish an energy consumption monitoring, reporting and analysis system. Introduce process analysis methods, system engineering principles, and the Plan-Do-Check-Act (PDCA) cycle management concept into the company's energy and carbon emission management. Establish a management system that covers the entire process of energy utilization and carbon emissions, and encourage various departments to establish a long-term mechanism for energy conservation and carbon reduction.
(2) Eliminate high energy-consuming electromechanical equipment: Sort out the existing high energy-consuming equipment, accelerate the elimination of high energy-consuming and obsolete electromechanical equipment, and continuously improve the energy efficiency of key energy-consuming equipment. Reduce the use of or, where technically feasible, replace the use of greenhouse gases of PFCs and HFCs, or use greenhouse gases with low global warming potentials to replace those with high global warming potentials.
(3) Transformation and application of energy-saving technologies: Apply lower-carbon production processes and technologies, transform production processes, reduce the use of fossil fuels and fossil materials in the production process, replace fossil fuels with electricity, and replace fossil raw materials with biomass, hydrogen, etc. Use advanced technologies to carry out energy-saving transformation of major energy-consuming systems, such as air-conditioning and ventilation systems, lighting systems, and power distribution systems.
(4) Utilization of renewable energy and low-carbon clean energy: Use energy with low carbon emission intensity to replace energy with high carbon emission intensity. For example, directly use electricity from renewable energy power plants such as wind power generation and photovoltaic power generation to replace electricity from the grid; use solar energy systems for heating or hot water to reduce electricity consumption, etc.
(5) Invest in the development of negative emission technologies and projects: Examples are afforestation or natural ecological restoration projects, to absorb CO_2 from the atmosphere through plant photosynthesis; and invest in CO_2 capture and storage projects.

2.4 *Achieving Carbon Neutrality Through Carbon Offsetting*

Carbon emission offsetting is a measure taken by an organization to deal with the remaining unavoidable carbon emissions after it has tried its best to take all feasible voluntary emission reduction measures. This strategy allows an organization to balance its own emissions and achieve carbon neutrality by supporting external emission reduction projects. These offsetting activities may include the following aspects.

- Afforestation: By increasing the green area, the photosynthesis of plants can be used to absorb CO_2 in the atmosphere, creating carbon sinks.
- Investment in renewable energy projects: By investing in solar energy, wind energy and other projects, we can encourage the development of clean energy and reduce the dependence on fossil fuels.
- Purchasing carbon credits: Purchase carbon credits through the carbon trading market to support projects that reduce greenhouse gas emissions, such as forest protection and methane emission reduction.

Through these offsetting measures, organizations can compensate for the carbon emissions that have not been reduced during production and operations, thus achieving the goal of net-zero emissions. This is an important way to promote global emission reductions and promote sustainable development. In this way, organizations can demonstrate their commitment to environmental responsibility and contribute to addressing climate change on a global scale. It needs to be emphasized again that the organization's approach to achieve carbon neutrality should be based on thorough consideration of scientific approaches. Carbon emission offsetting should still be used as a supplementary means for organizations to achieve carbon neutrality. After taking into account the characteristics of the organization, it can be an alternative plan after full implementation of direct emission reductions. Figure 2 shows a schematic diagram of corporate's carbon offsetting to achieve carbon neutrality.

The amount of carbon offsetting used by an organization to achieve carbon neutrality should generally follow the following principles.

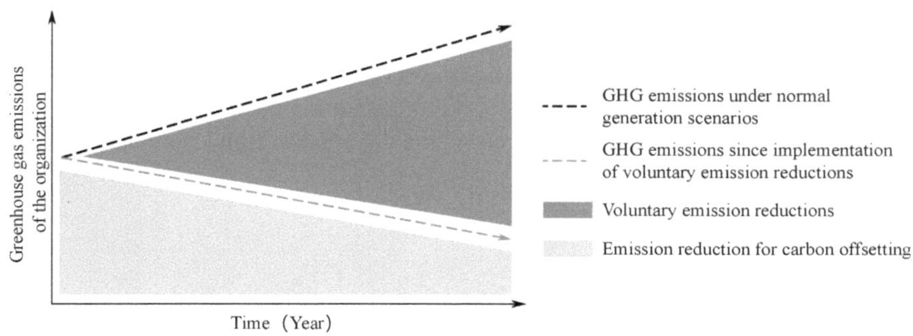

Fig. 2 Schematic diagram of corporate's carbon offsetting to achieve carbon neutrality

(1) Emission reduction for offsetting or carbon credits purchased should genuinely represent the additional greenhouse gas emissions reductions outside the organization's boundaries.
(2) The emission reduction projects used for carbon offsetting or carbon credits should meet the criteria of additionality, permanence and avoidance of double counting.
(3) Emission reduction for carbon offsetting should be certified by an independent third party.
(4) Emission reduction for carbon offsetting can only be issued after the emission reduction project has actually generated emission reductions.
(5) Emission reduction for carbon offsetting should be cancelled within a certain period of time after they have been used by the organization to achieve carbon neutrality, such as no more than 12 months.
(6) Carbon credits generated from carbon emission reduction projects should be supported by publicly available project documentation or a registry system, including project information, quantification methodology, and validation and verification procedures.
(7) Carbon credits generated by carbon emission reduction projects should be stored or cancelled in an independent and credible public registration system.

At present, common international and domestic carbon credit schemes include the Clean Development Mechanism (CDM), the Verified Carbon Standard (VCS), the Gold Standard (GS), the China Certified Emission Reduction (CCER), and the Puhui Certified Emission Reductions (PHCER). Organizations can purchase carbon credits to offset their remaining emissions. In addition to achieving carbon neutrality by offsetting the organization's remaining emissions with carbon credits generated by external emission reduction projects, organizations can also achieve their own net-zero emissions by investing in negative emission projects. Typical negative emission projects include forestry carbon sink projects, ocean carbon sink projects, and carbon capture and storage (CCS) projects. If the CO_2 absorbed from the atmosphere by the negative emission project equals to the organization's emissions, then the organization achieves carbon neutrality.

Investing in negative emission projects to achieve carbon neutrality is different from using carbon credits. Negative emission projects in which an organization invests should be included in the organization's overall accounting boundary, while emission reduction projects from which carbon credits are purchased must be outside the organization's accounting boundary. An organization cannot use carbon credits generated by emission reduction projects within the organization's boundary to offset its own carbon emissions.

3 Demonstration Cases of Carbon Neutrality at the Organizational Level

3.1 Apple's Carbon Neutrality Practices

In July 2020, Apple pledged on its official website to achieve full carbon neutrality by 2030, that is, to achieve "carbon neutrality" in its entire business operations, production supply chain and product life cycle. The company's initial target was to reduce carbon emissions by 75% from the 2015 level, and then deal with the remaining emissions by investing in carbon removal solutions.

According to Apple's carbon emission quantification results, the company's total emissions in 2019 were 25.1 million tonnes of CO_2 equivalent, of which 75% were emissions from the product manufacturing process, 16% from the product use process, and 6% from the upstream and downstream transportation of products, as shown in Fig. 3. The company's direct emissions come from sources such as diesel consumed by backup diesel generators, natural gas consumed in offices, propane, and fuel emissions from the fleet, accounting for less than 1%. The company's various sites and facilities use 100% renewable electricity, so the company's Scope 2 indirect emissions are 0.

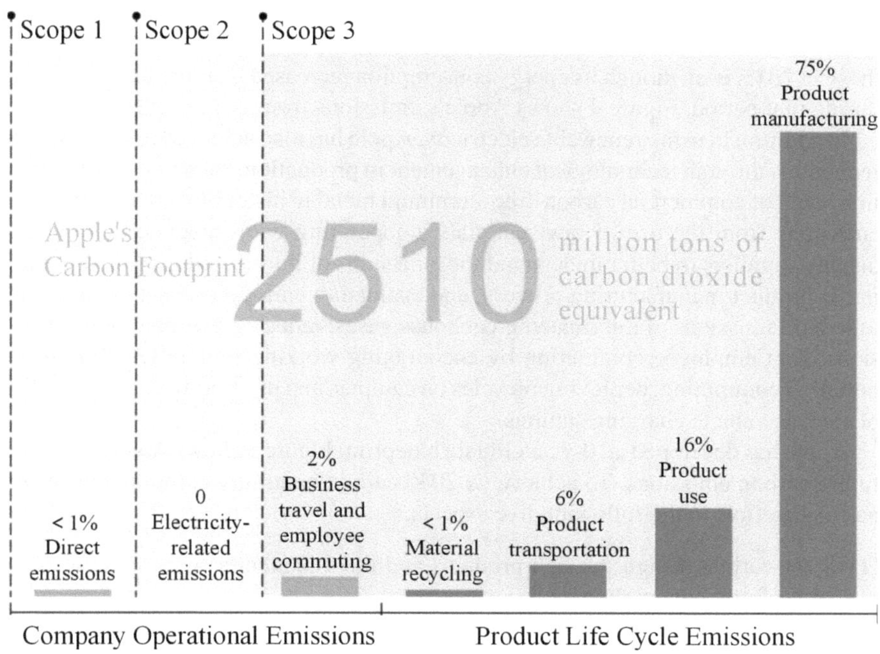

Fig. 3 Apple's 2019 carbon emissions quantification results. *Source* 2020 Environmental Progress Report

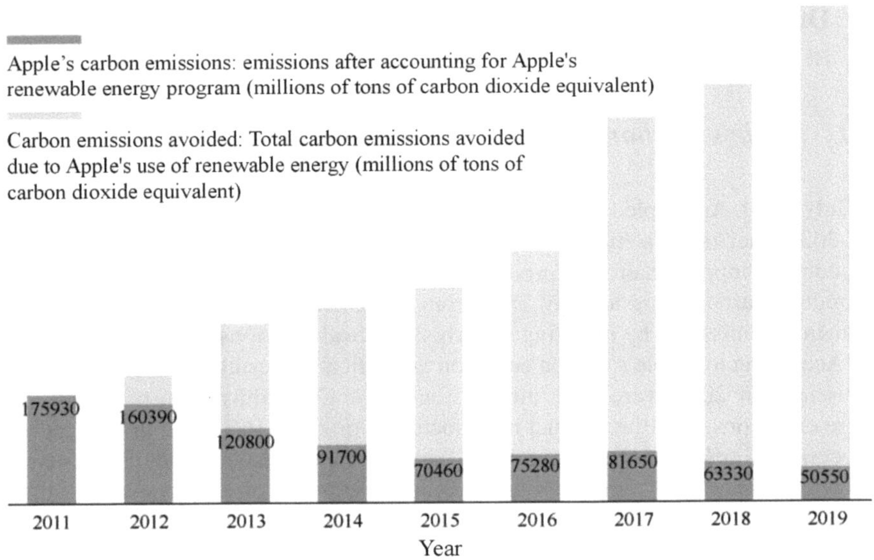

Fig. 4 Apple's emissions from 2011 to 2019. *Source* 2020 Environmental Progress Report

Since 2011, Apple has reduced its Scope 1 and 2 emissions by increasing its use of renewable electricity. In 2019, its Scope 1 and 2 emissions were 71% lower than those in 2011, even though its energy consumption increased by more than four times during that period. Figure 4 shows Apple's emissions from 2011 to 2019.

In addition to using renewable electricity, Apple has also achieved direct emission reductions through technological enhancement in production, such as purchasing the first batch of commercial carbon-free aluminum metal in history to reduce the carbon emissions from the use of raw materials; cooperating with integrated circuit and display suppliers to better understand the emissions of fluorinated greenhouse gases in the product manufacturing process, and evaluating emission reduction strategies to reduce emissions of fluorinated greenhouse gases; reducing greenhouse gas emissions from employee commuting by encouraging working remotely, public transportation commuting, deploying bicycles on campus, and promoting the construction of electric vehicle charging stations.

Apple has developed a 10-year climate blueprint. Figure 5 shows Apple's past and future carbon emissions. To achieve its 2030 carbon neutrality commitment, Apple has made effort in the following five aspects.

(1) Low-carbon design: Design products and the manufacturing process with the aim of reducing carbon emissions.
(2) Energy efficiency: Improve energy efficiency in company facilities and the supply chain.

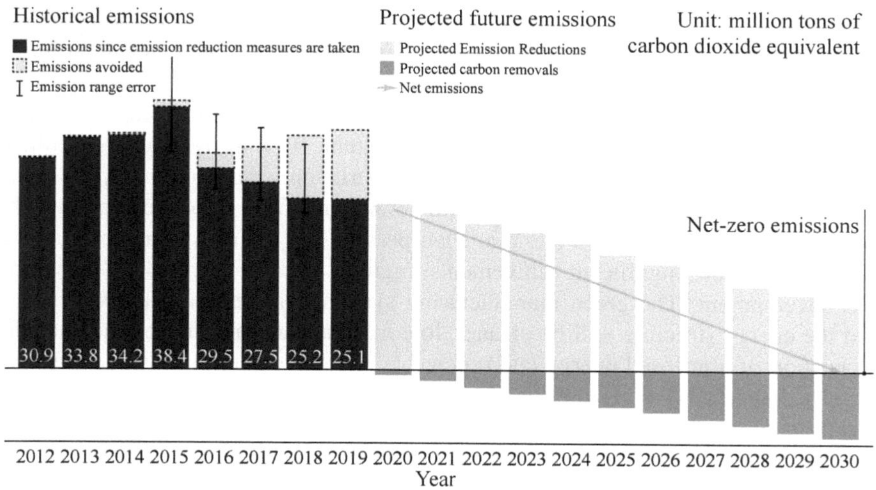

Fig. 5 Apple's past and future carbon emissions results. *Source* 2020 Environmental Progress Report

(3) Renewable electricity: Continue to maintain 100% use of renewable electricity at the company's facilities and promote the transition of the entire supply chain to 100% clean renewable electricity.
(4) Direct emission reduction: Avoid direct greenhouse gas emissions through technological solutions or by switching to non-fossil low-carbon fuels.
(5) Carbon removal: In parallel with emission reduction measures, increase investment in carbon removal projects, including natural solutions that can protect and restore global ecosystems.

In addition, Apple hires a third-party agency to verify the company's comprehensive carbon emissions every year to confirm the accuracy and stability of this quantification method. It will continuously review and optimize the company's carbon emissions to maximize the accuracy of carbon emission quantification and adjust the climate planning blueprint when necessary.

3.2 *Sichuan China Tobacco's Carbon Neutrality Practices*

a. Carbon Neutrality Commitments and Timetable

China Tobacco Sichuan Industrial Co., Ltd. (hereinafter referred to as Sichuan China Tobacco) responds to the country's call for green and low-carbon development, and adheres to the concept of green development. Through the formulation and issuance of a series of documents such as energy conservation, emission reduction, carbon reduction, carbon peaking and carbon neutrality action plans, Sichuan China Tobacco puts forward specific goals and action plans for energy conservation, emission reduction, carbon reduction and pollution prevention and control, and has established a clear timetable for carbon peaking and carbon neutrality.

During the 14th Five-Year Plan period, the focus will be on improving and enhancing the carbon emission management system. By improving energy utilization efficiency, optimizing energy consumption structure, and Building up an initial clean energy trunk transportation, a solid foundation for carbon peaking will be laid down. During the 15th Five-Year Plan period, it is planned to achieve carbon peaking and make significant achivements in green and low-carbon transformation and development. The green manufacturing system will be basically established, and the energy structure will be cleaner, low-carbon, safer and more efficient. The proportion of non-fossil energy and renewable energy consumption will be further increased, the energy consumption per 10,000 yuan of industrial added value will be greatly reduced, and the energy utilization efficiency will reach the internationally advanced level. At the same time, the green supply chain management will continue to be improved and the clean energy trunk transportation will be optimized. By 2060, a green, low-carbon circular development system and a clean, low-carbon, safe and efficient energy system will be fully established. The level of energy use efficiency and unit product energy consumption level will reach the advanced level in the industry, and the proportion of renewable energy consumption will be greatly increased, thus successfully achiving the carbon neutrality goal.

b. Quantification of Organizational Carbon Emissions

Quantification of organizational emissions is the foundation for achieving the goal of carbon neutrality. It requires organizations to quantify all relevant emissions by forming a detailed greenhouse gas emissions inventory. Through such a quantification process, organizations can clarify their own emission levels, identify potential areas and priorities for emission reduction, and provide data support for the formulation of effective emission reduction strategies. Sichuan China Tobacco commissioned a third-party agency, CEPREI Certification Body, to develop an annual organizational carbon emissions inventory in accordance with the internationally recognized standard and guide, ISO 14064-1:2018 Greenhouse gases—Part 1: Specification with guidance at the organization level for the quantification and reporting of greenhouse gas emissions and removals. This approach ensures the reliability, authenticity and fairness of the emission data, and provides a guarantee that the corporate's greenhouse gas emissions are accurately reported. According to the quantification results, Sichuan China Tobacco achieved a continuous decline in total carbon emissions between 2021 and 2023. The total carbon emissions in 2023 decreased by 3.62% compared with 2021. The trend of carbon emissions is shown in Fig. 6.

c. Voluntary Emission Reduction Actions

(i) Transformation of the Production Modes

Sichuan China Tobacco has taken a series of effective measures in the green disposal of tobacco waste to ensure that the waste is handled in an appropriate and environmental friendly manner, thereby mitigating its impact on the environment. The

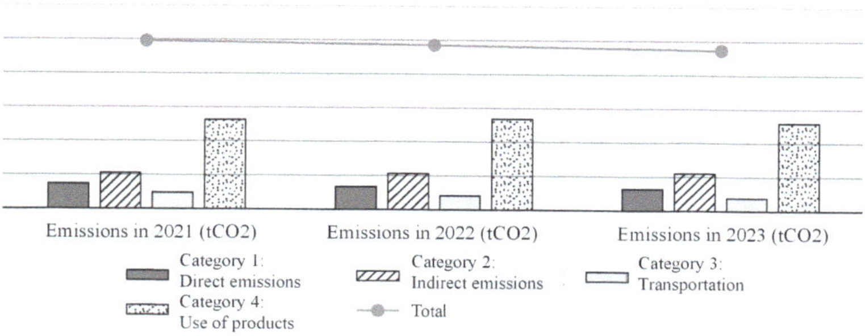

Fig. 6 The trend of carbon emissions of Sichuan China Tobacco from 2021 to 2023

company has implemented a strict classification and recycling system for waste generated during the production process. Through precise classification, different types of waste are treated appropriately to ensure that resources are recycled to the maximum extent. For hazardous waste, the company strictly abides by relevant laws and regulations to ensure its safe disposal and prevent potential harm to the environment and public health. Figure 7 shows the loading of special tobacco waste.

The company adheres to the principle of zero wastewater discharge. Through the construction and sharing of sewage treatment facilities, it continuously improves sewage treatment capacity and recycling rate of reclaimed water, achieving the comprehensive utilization of reclaimed water. Shifang Cigarette Factory shares the facilities of its sewage treatment station with Great Wall Cigarette Factory. The treated recycled water is used for greening the factory campus, watering the ground,

Fig. 7 Loading of special tobacco waste

Fig. 8 Wastewater treatment station of Chengdu Cigarette Factory

landscape watering, etc., which effectively reduces the dependence on fresh water resources and reduces the discharge of wastewater. Xichang Cigarette Factory uses treated reclaimed water for greening irrigation, which not only saves water resources, but also helps to improve the environmental quality of the factory premises, demonstrating the company's commitment to environmental protection. Suining Kuanzhai Printing uses treated wastewater for equipment cleaning, etc., thus realizing the recycling of wastewater, reducing environmental pollution, and improving the utilization efficiency of water resources. Chengdu Cigarette Factory has established a sewage treatment system of "pretreatment + biochemical treatment + filtration". About 1/3 of the treated reclaimed water is used for the irrigation of greens, about 1/3 is used for replenishing the deodorization system in the silk-making workshop, cleaning the garbage station, replenishing the cooling pool in the boiler room, etc. The remaining about 1/3 is used for cleaning municipal roads and municipal garbage transfer stations. Figure 8 shows the sewage treatment station of Chengdu Cigarette Factory, and Fig. 9 shows the reclaimed water treated by the sewage treatment station is being used for municipal road cleaning.

The company has taken measures to promote digital transformation, encourage changes in production methods, and deepen green manufacturing. The construction of a smart energy and environment management information system platform project has been launched, covering the company and its affiliated factories to achieve intelligent and information-based coordinated management. The platform features robust data analysis and forecasting/alerting functions. Through the analysis of indicator data, it can provide trend forecasts for energy consumption and pollutant emissions, allowing for timely identification of potential problems and risks. The platform incorporates the energy dual control + carbon emission dual control modules, setting and decomposing the energy and carbon control targets of each facility regularly. It can provide scientific data support for achieving dual control of energy and carbon emissions through data trend analysis and forecasting.

(ii) Transformation of Product Structure

The company has taken a series of innovative measures in green design to reduce resource consumption and environmental pollution in the production process. The

Fig. 9 Reclaimed water from the sewage treatment station being supplied for municipal road cleaning

company has successfully applied aluminum-free lining paper in three of its product brands, namely Blue Times, Tianxiaxiu (Gold), and Times Sunshine, as shown in Fig. 10, and it is expected to reduce the use of aluminum foil by 460 tonnes per year. By adjusting the water-based ink system, water-based ink has been applied in products such as Kuanzhai Guobao, Xiaoyao (Medium), Ge Diao (Slim), Ruyi (Slim), and Chunxiang (Medium), as shown in Fig. 11, reducing the use and emissions of oily pollutants. The company has carried out a weight reduction design for the packaging of "Pride" (Kuanzhai Ping'an, Medium) and "Pride" (Wuliang Nongxiang, Medium), as shown in Fig. 12, which simplifies the printing process at the printing factory. It not only reduces the use of paper and ink and lowers the production costs, but also reduces the generation of waste, which helps to improve resource use efficiency and reduce environmental pollution.

(iii) Transformation of Energy Structure

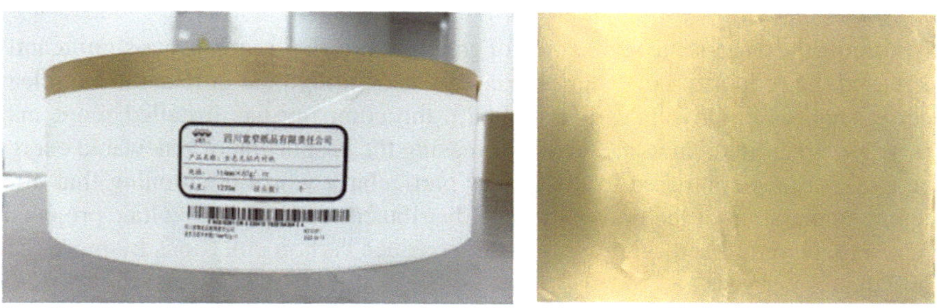

Fig. 10 Gold aluminum-free lining paper

Fig. 11 Water-based ink

Fig. 12 Weight reduction of packaging

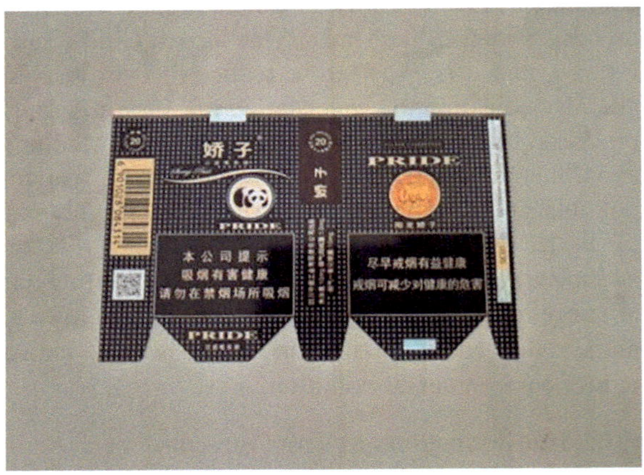

To promote the large-scale application of green energy, the company has significantly increased the proportion of renewable energy use through the efficient use of clean energy such as photovoltaics. As of 2023, the company has installed more than 800 solar street lights in its factories, increasing the proportion of renewable energy use by 3.5 times compared to 2020. It is particularly worth mentioning that after the implementation of the grid-connected distributed rooftop photovoltaic project of Xichang Cigarette Factory, it is expected to reduce carbon emissions by more than 1600 tonnes per year.

(iv) Transition to Green Logistics

Regarding the application of new energy vehicles in production logistics and transportation, three hydrogen fuel cell heavy trucks were bought. Taking advantage of their short refueling time, long driving distance, strong environmental performance and high environmental adaptability, they have been used for the business within the area of Chengdu, and the material transfer service has been extended to the downtown of Shifang City, Deyang City and Mianyang City and other routes. Two pure electric heavy trucks are being used to transfer materials between the warehouse areas of Shifang Cigarette Factory and between the warehouse areas and the production areas.

In terms of exploring low-carbon transportation methods for supply logistics, the company actively explores the use of low-carbon transportation methods such as railways and waterways for transporting flake tobacco. These transportation methods have the advantages of low cost, large transportation volume, and low environmental pollution. They can effectively reduce carbon emissions during transportation and are in line with the company's strategic goal of green and low-carbon development. By optimizing transportation routes and methods, the company not only reduces logistics costs, but also mitigates the impact on the environment.

(v) Transition to Green Supply Chain

For the "Pride" (Blue Era) cigarette products, the company conducted an experimental study on direct film wrapping technology, that is, no longer using the cigarette cartons but directly using transparent film for packaging, as shown in Fig. 13, which not only simplifies the packaging process, but also significantly reduces production costs. Furthermore, by improving the design of cigarette boxes, the reuse rate of recycled cigarette boxes has been successfully increased. This practice not only saves the use of new materials and reduces production costs, but also reduces the amount of waste generated, effectively promoting the transformation of packaging materials towards sustainable development. In addition, the company continues to expand the coverage of cigarette box recycling and cooperates with suppliers and distributors to establish an efficient recycling system.

d. Achieving Partial Carbon Neutrality Through Carbon Offsetting

After implementing possible voluntary emission reduction measures, carbon emission offsetting measures can be taken for the remaining unvoidable carbon emissions, in order to compensate for the carbon emissions that have not been reduced during production and operation, and achieve the net-zero emission target. Sichuan China Tobacco selected three high-end products in 2023—Kuanzhai Guobao Slim Cigarettes, Kuanzhai 1024 and Great Wall (GL NO. 1) cigars, and adopted low-carbon and environmentally friendly materials and processes in their production process. Through careful design and strict production management, these products reached high-quality standards. Under the premise of fully implementing voluntary emission reduction, the carbon emissions generated throughout the entire production life cycle were offset by purchasing carbon emission reductions. With this approach, these products achieved carbon neutrality at the product level, and obtained product carbon neutrality certification, becoming high-end zero-carbon products created by

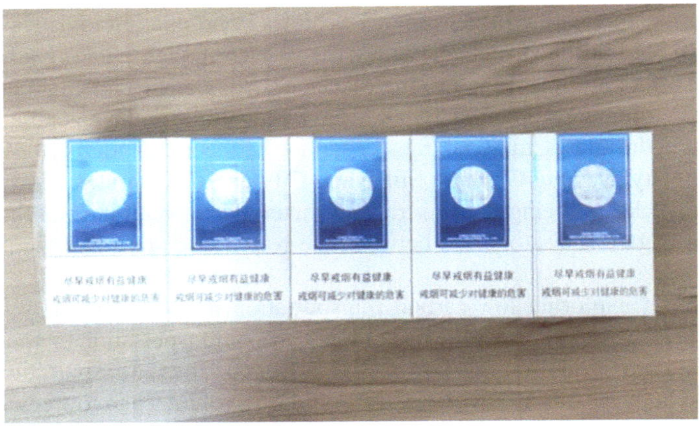

Fig. 13 "Pride" (Blue Era) direct film wrapping technology invented by Xichang Cigarette Factory

the company, as shown in Fig. 14. These zero-carbon products have been well accepted by the market for their unique design and high quality.

In addition, the company also held a "zero-carbon conference" through innovative planning to share and exchange the latest green and low-carbon achievements and experiences. In the process of holding the "zero-carbon conference", a series of measures were taken to reduce the carbon emissions of the conference itself. These measures include choosing a conference venue powered by renewable energy to ensure that low-carbon energy consumption during the conferenc, using electronic materials instead of paper printouts during the conference to reduce paper use and indirect carbon emissions, choosing low-carbon transportation and logistics solutions for necessary on-site activities, such as using electric vehicles or public Transportation, and employing carbon compensation strategies, such as purchasing carbon credits, to offset the carbon emissions generated by the conference.

Fig. 14 High-end zero-carbon products and carbon neutrality certificates

Chapter 5
Analysis of the Pathways to Carbon Neutrality at Product Level

1 Characteristics of Carbon Neutrality at Product Level

Achieving carbon neutrality for a product means reducing its carbon emissions to zero or close to zero through means such as reduction, offsetting or absorption. Carbon neutrality at the product level reflects a company's commitment to reducing greenhouse gas emissions throughout the entire life cycle of the product. Such carbon neutrality not only helps to reduce a company's impact on climate change, but also enables the company to reduce its negative impact on climate change, enhance its brand image, meet consumers' environmental protection requirements, and may obtain relevant tax reductions or incentives. Meanwhile, product carbon neutrality also helps to promote technological innovation and the development of renewable energy, and facilitates sustainable economic growth. Achieving product-level carbon neutrality usually involves assessing and managing the carbon emissions in each stage of the product, from raw material sourcing, production, transportation, use to final disposal, which helps to improve production processes, use low-carbon materials, optimize logistics, improve product energy efficiency, and increase the product recycling rate. In addition, companies may also need to use carbon offsetting projects, such as afforestation or purchasing carbon credits, to balance the carbon emissions that are difficult to eliminate through technological improvements. Through these comprehensive efforts, a product can achieve a net-zero increase in carbon emissions within its life cycle.

Carbon neutrality at the product level is a comprehensive process that covers the entire life cycle of a product, from raw material sourcing, manufacturing, transportation and distribution, consumer use to final disposal. This process requires assessing and managing the carbon emissions generated by the product in each stage to ensure that the overall carbon footprint is minimized. Achieving product carbon neutrality means reducing the impact of the product on climate change through various emission reduction measures, such as improving energy efficiency, using

C. Chen et al., *Sustainable Carbon Neutrality*,
https://doi.org/10.1007/978-981-95-1691-9_5

renewable energy, optimizing design, reducing material usage, improving production processes, extending product life, and enhancing recycling and reuse.

(1) Carbon neutrality at the product level is a company's active response to environmental protection requirements. It does not only help to reduce the product's contribution to climate change but also enhances the product's environmental protection attributes. As consumers' demand for sustainable products grows, companies can attract more consumers who pay attention to sustainable development by achieving product carbon neutrality and meet the market's expectations for environmental protection and social responsibility. By reducing the carbon emissions related to products, companies can reduce their negative impact on the environment. Meanwhile, through this process, companies can also promote their own technological innovation and improve the efficiency of resource utilization, thus driving the economy to develop in a more sustainable direction.
(2) Carbon neutrality at the product level is an important driving force for promoting a company's technological innovation. In order to reduce the carbon footprint of products and achieve carbon neutrality goals, companies must adopt or develop more efficient production technologies. This involves improving energy efficiency, using renewable energy, improving material usage, optimizing product design and improving manufacturing processes. These efforts not only help to reduce the energy consumption and carbon emissions of products during production and use, but also stimulate the development and innovation of new technologies, bring potential competitive advantages to companies, and promote technological progress in the entire industry. In this way, product carbon neutrality not only has a positive impact on the environment but also provides support for the long-term development and market adaptability of companies.
(3) Carbon neutrality at the product level is an important way for companies to adapt to the increasingly strict carbon emission policies and regulations around the world. By achieving carbon neutrality, companies can not only ensure that their products meet environmental protection standards, reduce the legal risks and potential costs faced due to violations of regulations, but also demonstrate their commitment to environmental responsibility and enhance the trust of the public and the market. In addition, many governments provide tax reductions, subsidies and other financial incentives to encourage companies to adopt green production measures. Through the practice of product carbon neutrality, companies can take advantage of these policy supports, reduce transformation costs and improve market competitiveness. Such policy-driven environmental protection actions not only contribute to the sustainable development of companies themselves but also have a positive impact on global emission reduction efforts and climate change responses.
(4) Carbon neutrality at the product level not only promotes the development and application of renewable energy technologies but also encourages the circular economy and the reuse of resources, thereby improving the efficiency of resource utilization. Achieving product carbon neutrality usually relies on renewable

energy, such as solar energy and wind energy, to reduce dependence on fossil fuels and lower greenhouse gas emissions. This transformation promotes the innovation and popularization of renewable energy technologies and accelerates the transition to clean energy. Meanwhile, product carbon neutrality also emphasizes improving resource efficiency and reducing waste in the production and consumption processes. This includes adopting the principles of the circular economy, such as reducing raw material usage, increasing material recycling, reuse and remanufacturing, and designing more durable and recyclable products. In this way, companies can reduce the demand for new resources, extend the service life of resources and lower the overall environmental impact.

(5) Carbon neutrality at the product level is a powerful branding and marketing strategy for companies. By achieving product carbon neutrality, companies not only demonstrate their commitment to sustainable development but also show their awareness of responsibility and leadership in environmental protection. Such commitment and actions help to build a positive brand image, enhance consumers' trust and loyalty to the brand. As consumers are increasingly concerned about companies' social responsibility and environmental impact, those companies that can provide carbon-neutral products are more likely to obtain market recognition and support. In addition, companies that achieve product carbon neutrality can highlight their environmental achievements in their marketing and communication strategies to attract consumers who are looking for sustainable products. Such a demonstration of transparency and responsibility can become an important basis for establishing long-term relationships between companies and consumers. Ultimately, product carbon neutrality not only helps companies to highlight their own advantages in a highly competitive market but also brings long-term economic and environmental benefits to companies.

2 Pathways for Implementation of Carbon Neutrality at Product Level

The entire process of product carbon neutrality can generally be divided into three stages. Firstly, quantify the carbon emissions of products. Quantify the carbon emissions at each stage of the product for the entire life cycle of the product. The greenhouse gas (GHG) emissions and removals generated during the raw material acquisition, production and processing, distribution, use and waste disposal stages of the selected product is calculated using the life cycle quantification method based on the ISO 14067:2018 *Greenhouse gases—Carbon footprint of products—Requirements and guidelines for quantification.* Secondly, plan the carbon reduction pathways. Each case of product carbon reduction is a process of realizing low-carbonization in the production, transportation and other links of its own industrial chain. Therefore, multiple ways can be adopted to plan the product carbon reduction pathways, and finally become a low-carbon solution for its own products. For example, in the production process, adopt energy-saving and emission reduction methods, such as photovoltaic power generation in factories, use of recycled plastic raw materials, etc., and reduce the product carbon footprint as much as possible through a series of measures. Thirdly, purchase carbon sinks for offsetting. For the carbon emissions that truly could not be avoided, the certified emission reductions should be cancelled, and eventually achieve carbon neutrality throughout the entire life cycle of the product. Generally speaking, many aspects need to be considered for product carbon neutrality, and a deep understanding of the entire life cycle of the product is required, and low-carbon and environmental protection should be practiced at each stage of the entire life cycle.

So, how exactly should product carbon neutrality be achieved? The implementation of product carbon neutrality includes organizational commitment and formulation of a carbon neutrality implementation plan, carbon emission accounting of the product life cycle, implementation of product emission reduction actions, offsetting of product carbon emissions, and evaluation of product carbon neutrality.

2.1 Organizational Commitment and Preparation of a Carbon Neutrality Implementation Plan

The carbon neutrality commitment and implementation plan of a company are the key steps in achieving sustainable development and responding to climate change. A company should predict the total amount of product carbon footprint and removals within the specified time boundary in the future according to the product functional unit and product system boundary, and formulate a product carbon neutrality implementation plan. The following is the detailed process for a company to formulate a carbon neutrality implementation plan.

(1) Carbon Neutrality Commitment: A company first needs to make a clear carbon neutrality commitment, which usually involves setting a target date by which the company will achieve carbon neutrality.
(2) Definition of Product Functional Unit: Determine the product functional unit, which is the standard unit for measuring the product service or benefit, such as the number of kilometers driven by each vehicle or the number of times each piece of clothing is worn.
(3) Determination of Product System Boundary: Clearly define the product system boundary, including the entire life cycle of the product from raw material acquisition, production, distribution, use to waste disposal.
(4) Carbon Footprint Prediction: Use the Life Cycle Assessment (LCA) method to predict the total carbon footprint of the product within the specified time boundary in the future, including direct emissions (such as energy consumption in the production process) and indirect emissions (such as emissions in the supply chain).

(5) Prediction of Carbon Removals: Evaluate and predict the carbon removals that can be achieved through natural carbon sinks (such as forests) or artificial carbon sinks (such as carbon capture, utilization and storage technologies) within the same time boundary.
(6) Setting of Carbon Neutrality Goals: Based on the predictions of carbon footprint and carbon removals, set specific carbon neutrality goals, including the total amount of carbon emission reduction and the specific timetable for achieving carbon neutrality.
(7) Carbon Footprint Reduction Strategy: Formulate strategies to reduce the product carbon footprint, which may include improving energy efficiency, using clean energy, improving production processes, using low-carbon materials, etc.
(8) Carbon Offsetting Strategy: If it is impossible to completely reduce carbon emissions, formulate a carbon offset strategy, including determining the amount of carbon that needs to be offset, selecting offset projects (such as afforestation, carbon capture, utilization and storage projects) and purchasing carbon credits.
(9) Preparation of Implementation Plan: Plan the implementation steps in detail, including the technology roadmap, timeline, responsibility allocation, resource allocation and budget plan.
(10) Monitoring, Reporting and Verification: Establish a monitoring, reporting and verification (MRV) system, regularly track the progress of carbon neutrality, and report the results to stakeholders, which may involve third-party verification to ensure transparency and credibility.
(11) Risk Management: Identify and evaluate the risks that may be encountered during the implementation process, such as technological changes, market fluctuations, policy changes, etc., and identify corresponding risk management measures.
(12) Stakeholder Support: Communicate with stakeholders such as employees, customers, suppliers, investors and communities to obtain their support and jointly promote the achievement of carbon neutrality goals.
(13) Continuous Improvement and Innovation: Encourage continuous improvement and innovation to adapt to the constantly changing environmental and technological conditions and improve the effectiveness of the carbon neutrality implementation plan.
(14) Policy and Market Adaptability: Consider the impact of policies and market mechanisms on the carbon neutrality implementation plan to ensure the adaptability and flexibility of the plan.

Through these steps, a company can formulate a comprehensive and systematic carbon neutrality implementation plan, which not only helps the company achieve its own carbon neutrality goals but also contributes to the global response to climate change.

2.2 Carbon Emission Accounting in the Product Life Cycle

Carbon footprint is an indicator used to measure the amount of carbon dioxide emissions directly or indirectly caused by individuals, organizations, products or countries within a certain period of time. Carbon footprint can be classified as national carbon footprint, urban carbon footprint, organizational carbon footprint, corporate carbon footprint, household carbon footprint, product carbon footprint and personal carbon footprint according to its application level. Among them, product carbon footprint is the most widely used concept. Product carbon footprint is a type of carbon emission accounting and generally refers to the total amount of carbon emissions generated in the processes such as raw material acquisition, transportation, production and factory sales of products. It is an important indicator for measuring the green and low-carbon level of manufacturing organizations and products. The product carbon footprint records the results of product carbon footprint quantification in terms of carbon dioxide equivalent per functional unit. The product functional unit should be clearly defined, clear and measurable, and the functional unit should be consistent with the evaluation objectives and contents. The functional unit provides relevant references for output and input.

In recent years, some countries have gradually established accounting, evaluation and certification systems for the carbon footprint of key products. More and more multinational companies have also incorporated product carbon footprint into the requirements for sustainable supply chain management. In 2023, five departments including the National Development and Reform Commission jointly issued the "Opinions on Accelerating the Establishment of a Product Carbon Footprint Management System". In 2024, multiple departments including the Ministry of Ecology and Environment issued the "Implementation Plan for Establishing a Carbon Footprint Management System", systematically deploying key tasks and building the overall framework of the product carbon footprint management system. It was proposed to start from the product carbon footprint, improve domestic rules, promote international convergence, and establish a unified and standardized carbon footprint management system.

(1) The Concept of Product Carbon Footprint

Product carbon footprint refers to the total amount of greenhouse gases released or absorbed by a certain product during its life cycle process, that is, the accumulation of various greenhouse gas emissions or absorptions in multiple stages from raw material acquisition, product production (or service provision), transportation, use to final recycling and disposal. Product carbon footprint has become an effective quantitative indicator for measuring the performance, management level of companies and the impact of products on climate change.

The accounting of product carbon footprint adopts the Life Cycle Assessment method (LCA). As a systematic method, LCA covers all stages from raw material acquisition, production, transportation, use to waste disposal and recycling. By evaluating the environmental burden caused by products throughout the life cycle (from raw material acquisition to waste disposal), it realizes the control of the whole process of carbon footprint. The LCA, as an environmental management tool, has gone through years of development. It was officially proposed by the Society of Environmental Toxicology and Chemistry (SETAC) in 1990 and has been widely used in many European and American countries. In 2003, the European Commission identified LCA as the best framework for evaluating the potential environmental impacts of products. In 2009, the European Union issued the Ecodesign Directive, requiring companies to pay attention to the use of resources in all stages of the product life cycle. In 2023, the New Battery Regulation of the European Union officially came into effect, mandating all battery companies entering into the EU market to use the Product Environmental Footprint (PEF) method under the LCA system to calculate the product carbon footprint.

The LCA method is a standardized method for systematically and quantitatively evaluating the resource and environmental efficiency in the product life cycle process. It helps producers identify the production stage where environmental problems occur through tracing the upstream and downstream production and consumption processes of products, and further avoid the transfer of thsoe problems between different life cycle stages of products and different types of environmental impacts. According to the differences in system boundary setting and model principles of the methods, the currently commonly used life cycle assessment methods can be divided into three categories: Process-based LCA (PLCA), Input–output LCA (I-OLCA), and Hybrid-LCA (HLCA). Process-based LCA is currently the mainstream evaluation method. For the calculation of carbon footprint at the micro level (specific products or services), the process life cycle method is generally adopted. Input–output LCA overcomes the drawbacks of boundary setting and inventory analysis in process life cycle assessment and introduces the economic input–output table. This method is generally applicable to calculations at the macro level (such as countries, sectors, companies, etc.). Hybrid-LCA is a life cycle assessment method that combines process life cycle assessment and input–output life cycle assessment. The advantage of this method is that it can not only avoid truncation errors but also evaluate specific products and their entire life cycle stages (use and disposal stages) in a more targeted manner. However, due to its high standards for data and calculation and the difficulty in building models, it has not been widely practiced yet.

The implementation of LCA and carbon footprint certification helps to maximize resource conservation and greenhouse gas emission reduction, which is of great value and significance for the green development of industries, the upgrading and transformation of industries, and dealing with potential trade barriers in exports. The product carbon footprint certification is based on the standard ISO 14067:2018 *Greenhouse gases—Carbon footprint of products—Requirements and guidelines for quantification*, which defines specific requirements for the certification scope, certification principles, quantification methods for life cycle greenhouse gas emissions and the contents of carbon footprint accounting reports of product carbon footprint certification.

The values of implementing product carbon footprint certification mainly include:

- Obtaining the life cycle carbon footprint indicator results of products, which can be used to compare the carbon emission situations of products under different processes among companies of the same type and select more environmentally friendly process technologies.
- The carbon footprint results can be used for downstream product design and low-carbon management of the supply chain to promote the low-carbon development of the entire industrial chain.
- The carbon footprint results can be used for market publicity to show the advantages of products and/or production processes in terms of carbon emissions and provide a basis for low-carbon choices of downstream manufacturers or end consumers.

(2) Product Carbon Footprint Certification Process

When a company calculates the carbon footprint of its products, it needs to identify all sources and types of greenhouse gas emissions and use emission factors or other forms to calculate the emissions throughout the life cycle. In cases where the production of other products is involved in the product production process (such as the production of co-products), the physical allocation method shall be preferentially used to allocate the carbon footprint. When physical allocation is not possible, the economic allocation method shall be adopted. For situations involving greenhouse gas removals, the accounting of the removal amount shall comply with public standards or guidelines. Ultimately, the evaluation report of the product carbon footprint can be issued by a qualified third-party body to ensure the objectivity and authority of the evaluation. This series of accounting and evaluation processes helps companies accurately grasp the impact of products on climate change and take effective measures to promote sustainable development.

a. Determine the Certification Boundary and Functional Unit

Product carbon footprint certification is the accounting of the carbon footprint of a specific model of a manufacturer's product. Therefore, it is first necessary to determine the certification boundary, that is, the manufacturer of the product and the specific product name, series, specification, and model. Applications for certification for products of different models must be made separately due to differences in their raw and auxiliary materials and energy consumption in the production process.

b. Identify the Product Emission Process

When conducting product carbon footprint accounting, it is necessary to ensure that the setting of the system boundary is comprehensive and reasonable. When determining the product system boundary, an appropriate boundary type should be selected based on the principle of the product life cycle. When determining the carbon footprint certification of a specific product, the boundary of the life cycle system needs to be determined according to factors such as the purpose of certification, product attributes, and data availability. Common system boundaries include "cradle to gate" and "cradle to grave". Usually, "cradle to grave" is applicable to end products, covering the entire life cycle from raw material acquisition to recycling or disposal after product use; while "cradle to gate" is applicable to intermediate products, mainly focusing on the process from raw material acquisition to product leaving the factory. Selecting an appropriate system boundary is crucial for ensuring the comprehensiveness and accuracy of carbon footprint accounting. All emission processes with expected emissions accounting for 1% or more of the product carbon footprint should be taken into consideration to ensure the accuracy of accounting. Any emissions not included should be controlled within 5% of the total product carbon footprint, and for these excluded emissions, there must be records in writing and justification.

The identification of the emission process should comprehensively cover all stages of the product life cycle, including raw material acquisition, energy consumption, manufacturing, facility operation, transportation and storage, product use, and end-of-life treatment of the product. It is worth noting that the embodied emissions of production equipment are usually not included. Meanwhile, human input, employee commuting, transportation provided by human and animal power, and consumers' transportation to and from sales points should also not be included in the product carbon footprint.

During the carbon footprint certification process, the life cycle system boundary of the product is determined by the certification client based on specific circumstances. Generally speaking, intermediate products or non-energy-consuming products are more suitable for choosing the "cradle-to-gate" boundary, while end products or energy-consuming products tend to choose the "cradle-to-grave" boundary to comprehensively evaluate the potential impact of products on climate change. With this method, it can be ensured that the accounting of the product carbon footprint is both comprehensive and targeted, helping companies take effective measures to reduce carbon emissions and promote the achievement of carbon neutrality goals.

c. Collect Life Cycle Assessment Data

Product carbon footprint accounting is a comprehensive process that collects and organizes key data based on the LCA method. These data include the following aspects:

- Consumption of raw and auxiliary materials: The types and quantities of raw materials and auxiliary materials required for product production.
- Supplier information and transportation distance: Information on suppliers of major raw and auxiliary materials and the distance for transporting materials to the factory.
- Energy consumption in the production process: The types and quantities of energy used in the production process.
- Consumption of packaging materials: The types and quantities of materials used for product packaging.
- Energy and greenhouse gas consumption in the product use stage: Energy consumption and greenhouse gas emissions during product use.
- Energy consumption in the product recycling and disposal process: Energy use during product recycling or disposal.

- Quantity of disposal: The disposal methods and quantities after product disposal.

This data mainly comes from the bill of materials (BOM table) of the product, production reports, etc. For data that cannot be directly measured, reasonable estimation methods should be adopted following the industrial standards or specifications. To ensure the accuracy and credibility of the carbon footprint accounting results, the verification team will conduct a strict quality assessment of the collected data. The dimensions of the assessment include the following aspects:

- Representativeness: Whether the data can accurately represent the actual situation of the product carbon footprint.
- Completeness: Whether the data comprehensively covers all stages of the product life cycle.
- Reliability: Whether the data source is reliable and whether the data collection and processing methods are scientific.
- Consistency: Whether the data remains consistent under different times and conditions.

Through such meticulous data collection and quality assessment, it is ensured that the accounting results of the product carbon footprint are both accurate and credible, providing valuable information for companies, helping them optimize the production process··, reduce carbon emissions, and show the environmental protection advantages of their products in the market.

d. Calculate the Product Carbon Footprint

Calculating the product carbon footprint is a precise and systematic process that requires the use of the highest-quality data to ensure the accuracy and credibility of the results. In the data collection stage, the measured emission factor data shall be preferentially adopted, which can provide the most direct and accurate estimation of emissions. If the measured data is unavailable, local, national and international public data sources should be considered in turn to ensure the relevance and reliability of the data.

After completing the collection and collation of data, a dedicated accounting model is established for the specified product. This model will include all relevant accounting units and the collected data will be entered into the system. Then, a clear calculation plan is determined, which usually involves selecting appropriate emission factors and calculation methods for adherence to the principles and standards of life cycle assessment. The result will reflect the total amount of greenhouse gas emissions throughout the entire life cycle of the product from raw material acquisition, production, use to final disposal.

The accuracy and transparency of the accounting process are crucial for obtaining accurate carbon footprint data. Therefore, all calculations and data sources should be recorded and archived in detail for subsequent verification. In this way, companies can ensure that the accounting results of their product carbon footprint are both scientific and fair, helping them demonstrate their commitment to environmental responsibility in the market and take actions to reduce the impact of their products on climate change.

e. Prepare the Report and Issue the Certificate

Report preparing is a key link in the product carbon footprint accounting process, ensuring the formality of the accounting results. Generally, the certification audit team prepares a complete product LCA report based on the data. Figure 1 shows the product carbon footprint certificate. The content of the certificate usually includes:

- Product carbon footprint accounting information table, which lists all relevant emission data in detail.

CEPREI

No: CEPREI-2025-CFP-0058
Issue date: 2025.06.27
Expiration date: 2027.06.26

Wang Chunhui
General Manager
CEPREI CERTIFICATION BODY

Carbon Footprint of Products
Verification Statement

(Original)
This is to certify that

The results of the quantification of the carbon footprint of

XX Product

has been verified in accordance with **ISO 14064-3: 2019** as meeting the requirements of
ISO 14067: 2018 and GB/T 24067-2024

Client: XX Co Ltd
Product Producer(Manufacturer):
ManufacturingPlant:
ProductionAddress:

Product Series / Specifications / Models: 2.0mm
System Boundary: Raw material acquisition stage Production stage
Time Boundary for Data: 2024.01.01-2024.12.31
Functional Unit: 1 ton Heat strengthened patterned glass (2.0mm)
CFP Result (CO_2-eq): 1.28 t
Level of Assurance: Reasonable assurance
Materiality: 5%

1.28 t
CO_2
CEPREI

Fig. 1 Product carbon footprint certificate

- Definition of accounting objectives and scope, clarifying the purpose and boundary of accounting.
- Data collection process and summary, showing the source and processing method of data.
- Product carbon footprint results and analysis, explaining the calculation results of the carbon footprint and related factors.
- Life cycle interpretation, elaborating on the selection and reasons for the product system boundary.
- Conclusions and suggestions, providing suggestions and improvement directions based on the accounting results.

2.3 *Implement Product Emission Reduction Actions*

Implementing product emission reduction actions is a crucial step for companies to achieve the carbon neutrality goal. Companies can adopt reasonable greenhouse gas emission reduction strategies based on their actual situations and determine the effectiveness of these actions. Implement product carbon footprint reduction actions based on the implementation plan to achieve the expected carbon footprint reduction targets to a maximized level. The following are the steps for companies to effectively implement product emission reduction actions.

- Formulate Emission Reduction Strategies: Based on the data results of product carbon footprint, formulate reasonable greenhouse gas emission reduction strategies, such as improving production processes, using more energy-efficient equipment, optimizing supply chain management, and usingclean energy.
- Determine Action Effectiveness: Before implementing emission reduction actions, predict and determine the effectiveness of these actions, including quantified emission reduction expected and possible cost-effectiveness.
- Develop an Implementation Plan: Following the emission reduction strategies, develop a detailed implementation plan, including specific action steps, timetables, responsibility assignments, and required resources.
- Implement Emission Reduction Actions: In accordance with the implementation plan, take specific emission reduction actions, such as technological upgrades, process optimizations, and material substitutions.
- Monitor and Track Progress: During the implementation, continuously monitor and track the effects of emission reduction actions to ensure they are aligned with the expected goals. Conduct regular performance evaluations, including quantitative analysis of emission reduction effects and cost-effectiveness evaluations.

Through these steps, companies can ensure the effectiveness of their product emission reduction actions and achieve the expected carbon footprint reduction targets to a maximized level.

2.4 Product Carbon Emission Offsetting

Greenhouse gas emission offsetting is a practice of compensating for the carbon emissions that companies themselves cannot reduce by purchasing carbon credits. Companies can use carbon credits from the following types of projects to achieve product carbon neutrality: project carbon credits issued by CCER, CDM, SDM, GS, VCS, etc. For the greenhouse gas emissions generated by companies outside China, it is recommended to give priority to carbon credits issued by international emission reduction mechanisms for offsetting, which are generated from projects in the country where the company is located; or carbon credits generated from projects under local carbon emission reduction mechanisms recognized by the competent local authroity of the company's location. Offsetting the carbon footprint of products produced in a specific time period by obtaining carbon credits, and the time for accounting shall not be earlier than the year of product neutralization. Once carbon credits are used for product carbon neutrality, they should be cancelled with the corresponding management agency, and companies should keep relevant cancellation certificates for at least 5 years. The following are the common steps for companies to offset the carbon footprint of products produced in a specific time period by obtaining carbon credits.

1. **Carbon Credit Market Research**: Research the carbon credit market, understand different types of carbon credits, their sources, prices, quality, certification standards, etc.

   Carbon credit market research is a comprehensive and in-depth analysis work, which requires meticulous research on multiple dimensions of the market. First, understanding different types of carbon credits is the first step. These credits may come from different emission reduction projects, such as renewable energy projects, energy efficiency improvements, forest carbon sinks, methane reduction, etc. Each type of carbon credit has its specific generation mechanism and application scenarios.

   The source is the key to the identity of carbon credits. It involves the specific geographical location of the project, the project type, and the entity implementing the emission reduction activities. For example, some carbon credits may come from clean energy projects in developing countries, while others may originate from industrial process improvements in industrialized countries. The price is an important indicator to measure the vitality of the carbon credit market. It is affected by various factors such as market supply and demand, policy orientation, and economic conditions. The trading price of carbon credits not only reflects their value as a commodity but also serves as a barometer of the market's recognition of emission reduction actions. Quality is the core of the carbon credit market. It is directly related to the authenticity, effectiveness, and sustainability of carbon emission reduction projects. High-quality carbon credits should ensure the additionality of their emission reduction, avoid double-counting, be permanent, and be verified by strict third-parties. These standards are usually evaluated and confirmed by certification bodies according to a series of predefined standards and methods. Certification standards are the rules and procedures to ensure the quality of carbon credits. They provide norms for the generation, issuance, and trading of carbon credits. Certification bodies such as Verra and Gold Standard ensure the international recognition and market acceptance of carbon credits by formulating detailed project methodologies and audit processes.

Overall, carbon credit market research requires a systematic analysis of market size, growth trends, price fluctuations, supply-demand dynamics, policy environment, technological development, certification mechanisms, etc. In-depth analysis of these complex factors can provide decision-making support for market participants, promote the healthy development of the carbon market, and advance global climate action.

2. **Select Appropriate Carbon Credit Projects**: Based on the research results, select carbon credit projects that are in line with the company's values and sustainable development goals, such as renewable energy projects, forest carbon sink projects, etc. Selecting an appropriate carbon credit project is a multi-dimensional decision-making process. It requires companies to conduct detailed analysis and evaluation based on a deep understanding of the market status, clear goals, and values. Selecting an appropriate carbon credit project is a strategic decision making process. It requires companies to make a wise investment choice based on a comprehensive consideration of their own values, sustainable development goals, and market research results. This process not only involves a deep understanding of the carbon credit market but also an accurate grasp of the company's own needs and long-term goals.

   First, companies need to screen carbon credit projects according to their own values. Usually, choose projects that are consistent with the company's corporate social responsibility (CSR) strategy, such as in aspects of biodiversity protection or environmental improvement. For example, if a company values ecological protection, it may prefer to choose forest carbon sink or nature reserve restoration projects. Second, companies should consider their sustainable development goals and ensure that the selected projects can support the realization of these goals, including reducing their own carbon footprint, improving energy efficiency, or promoting low-carbon practices in the supply chain. Choosing projects that can help companies achieve these goals can enhance the company's competitiveness and market position in terms of sustainable development. Based on market research, companies need to evaluate the potential value and risks of different carbon credit projects, including a comprehensive analysis of the project's environmental benefits, social impact, technical feasibility, economic returns, and compatibility with existing policies and market mechanisms. For example, companies may consider whether the project can provide long-term and stable carbon emission reductions, and whether these emission reductions can be verified and certified by a third-party. In addition, when choosing carbon credit projects, companies should also consider their impact on the company's reputation and brand image. Choosing projects with high transparency and good governance structures can help companies avoid risks related to "greenwashing" or false propaganda. Finally, when choosing carbon credit projects, companies should also consider their flexibility and scalability. Choosing projects that can be adjusted as the company's scale and business needs change can ensure that the company remains competitive in a changing market environment.

3. **Ensure the Additionality of Carbon Credits**: Ensure that the purchased carbon credits come from projects that can provide additional carbon emissions reduction or carbon sinks, and avoid double-counting or reuse. Ensuring the additionality of carbon credits is a core element in carbon credit market research and purchase decision-making. Additionality means that the emission reduction amount generated by a carbon emission reduction or carbon sink project would not occur otherwise except for the project itself, that is, these emission reduction amounts are generated additionally, rather than simply attributing existing emission reduction achievements to the project. Additionality is one of the key indicators for evaluating the quality of carbon credits. It ensures that the environmental benefits represented by carbon credits are real and new, rather than double-counting of existing emission reduction measures. This is crucial for maintaining the integrity and efficiency of the carbon market and also protects the company's reputation, avoiding the risk of "greenwashing".

2.5 *Product Carbon Neutrality Evaluation*

Conducting product carbon neutrality evaluation is a systematic task. It requires a quantitative assessment of the product's carbon footprint within a specific time period, including the total carbon footprint, carbon emission reduction, and the achievement of carbon neutrality. This process first requires determining the greenhouse gas emissions generated by the product at various stages such as production, use, and disposal, and then reducing these emissions through the implementation of various emission reduction measures. When the total carbon footprint of a product within its life cycle does not exceed the carbon credits used by the company for offsetting, the product can be considered to have achieved carbon neutrality. This determination is based on rigorous standards and procedures to ensure that the carbon credits used for offsetting are unique and will not be used for any other purposes.

When implementing carbon neutrality evaluation, companies should entrust a third-party organization with corresponding qualifications and professional capabilities to conduct it to ensure the impartiality and accuracy of the evaluation. The third-party organization will account for and verify the product carbon footprint based on internationally recognized standards and methods, such as ISO 14064. The evaluation needs to be consistent with the company's carbon neutrality implementation plan to ensure that the evaluation results can truly reflect the achievement of product carbon neutrality.

In addition, when conducting carbon neutrality evaluation, companies should also consider all aspects of the product's full life cycle, including raw material acquisition, production and processing, packaging and transportation, use process, and final disposal. By comprehensively analyzing the carbon emissions of these links, companies can have a more complete understanding of the product's environmental impact and take effective measures for emission reduction.

During the carbon neutrality evaluation process, companies should also pay attention to the long-term effects and sustainability of carbon emission reduction. In addition to achieving short-term carbon neutrality goals by purchasing carbon credits, companies should also actively explore and implement long-term emission reduction strategies, such as improving energy efficiency, using clean energy, and optimizing product design. These measures not only help reduce greenhouse gas emissions but also enhance the competitiveness and market position of the company.

In summary, product carbon neutrality evaluation is a complex process involving multiple steps and aspects. Companies need to ensure the accuracy and reliability of the evaluation results through scientific methods and strict procedures. At the same time, companies should continuously optimize the carbon neutrality implementation plan and improve the effectiveness of emission reduction measures to achieve true environmental sustainability.

3 Product Carbon Neutrality Examples

3.1 Apple's Carbon-Neutral Products

Apple claims that it achieved carbon neutrality in its corporate operations (Scope 1 and Scope 2 emissions) in 2020 through means such as using clean electricity, and setting a more ambitious goal to achieve product carbon neutrality by 2023, that is, products will achieve net-zero carbon emissions throughout their life cycle. Apple CEO Tim Cook said at the China Development Forum 2024 Annual Meeting that Apple plans to achieve carbon neutrality for all its products by 2030. In the three aspects of raw materials, production, and transportation, Apple is promoting hundreds of carbon-reduction related projects. Cook emphasized that without the support of suppliers, Apple will not be able to achieve its carbon neutrality goal, nor will it be possible to launch a carbon-neutral watch. More than 200 of Apple's suppliers worldwide have committed to using 100% renewable energy to produce Apple products, including 55 Chinese suppliers such as BOE and Luxshare Precision. In 2023, Apple grandly launched its first carbon–neutral product-Apple Watch Series 9 at the autumn launch event, as shown in Fig. 2. The emissions throughout the life cycle of a product come from the raw materials and auxiliary materials used in the product, the energy consumption in the product production process, product transportation, use, etc. How did Apple achieve net-zero emissions throughout the life cycle?

1. Step 1: Understand the Product Emission Status

Fig. 2 Apple Inc.'s carbon-neutral product—Apple Watch Series 9

Based on the situation without emission reduction: Without using clean electricity, with the upstream emission intensity of purchased raw materials and auxiliary materials, and the product transportation methods from 2017 to 2019, Apple used the LCA method to evaluate the carbon footprint of Apple Watch Series 9 throughout its life cycle as 36.7 kg of carbon dioxide equivalent, among which the electricity emissions in the product and raw material production process is accounted as the largest contributor.

2. Step 2: Carbon Reduction throughout the Life Cycle

Apple reduced the carbon footprint of Apple Watch Series 9 by 78% through carbon reduction in all links of the product's life cycle.

(1) Upstream Raw Materials and Production Links: The product used more than 30% recycled and regenerated raw materials.

The watch case is made of 100% recycled aluminum metal.

Gold in multiple printed circuit board coatings, tin in solder, rare-earth elements in magnets, tungsten in the Taptic Engine, and copper foil in the motherboard, etc. are made of 100% recycled materials.

The battery uses 100% recycled cobalt.

The watch strap uses 82% recycled yarn, and some of its raw materials come from discarded fishing nets.

Renewable plastics are used in the speaker, and the proportion of recycled plastics used in the product reaches 25%.

(2) Parts and Product Production Links

Use Clean Electricity: Carry out the Supplier Clean Energy Program to promote 100% of the suppliers related to the production of this product to use clean electricity. In addition, as of March 2023, more than 250 suppliers have joined the Supplier Clean Energy Project, accounting for more than 85% of Apple's direct manufacturing expenditure.

Improve Manufacturing Efficiency: None of the assembly plants related to this product generate waste that needs to be landfilled.

Reduce Plastic Packaging: The product packaging is made from 100% fiber materials, contains no plastics except for inks, coatings, and adhesives, and includes 47% recycled materials.

(3) Product Transportation Link

Reduce Packaging: The product packaging box has been redesigned. It is smaller and more efficient than the previous-generation box, with a 23% reduction in volume, improving product transportation efficiency.

Reduce Air Transportation: Give priority to less carbon-intensive transportation methods, such as rail and sea. More than 50% of carbon-neutral products use non-air transportation methods within their life cycle.

(4) Product Use and Disposal Links

Reduce User Charging Emissions: Invest in renewable energy projects such as the IPRadian solar project in Brown County, Texas, USA, striving to 100% offset the estimated electricity consumption required for users to charge the carbon-neutral Apple Watch.

Improve Product Power-Use Efficiency: Use software and energy-saving components to intelligently manage power consumption.

Prolong Product Life: Use durable materials, with a water-resistance level of 50 m and a dust-proof level of IP6X.

Maximize Product Recycling: Apple launched the Apple Trade In program, where customers return their old devices and accessories to Apple. Eligible devices can be exchanged for credit cards or Apple Store gift cards. Apple also created the Apple Recycler Guide to provide guidance for professional electronics recyclers on how to safely disassemble Apple products to maximize resource recovery.

3. Step 3: Carbon Offset

After reducing the product carbon footprint through emission reduction measures, Apple purchases credit quotas generated by projects that meet international standards [such as Verra, Climate, Community and Biodiversity (CCB) standards, and Forest Stewardship Council (FSC)] to offset the remaining emissions and achieve product carbon neutrality.

From the pathway of Apple's product carbon neutrality, taking emission reduction measures throughout the product life cycle to reduce product carbon emissions is an important measure and necessary pathway to create a carbon–neutral product. Taking carbon offsets under the condition of minimizing product carbon emissions is the truly meaningful carbon neutrality.

3.2 POP MART's Carbon-Neutral Products

In September 2023, POP MART released its first "carbon-neutral" trendy toy product, the DIMOO X Mongolian Beaver hand-painted figure. As the first "carbon-neutral" product in the trendy toy industry, this product not only represents the company's commitment to environmental protection and sustainable development but also demonstrates its innovation capabilities in product design, supply chain management, and logistics optimization. Through a series of carbon-reduction measures, the product achieved carbon neutrality throughout its life cycle from raw material procurement to production, logistics, and sales, providing exploration and practice for the green transformation of the industry.

First, in the product design stage, the environmental protection concept was integrated into it. The product used 20% recycled materials for the first time, reducing the demand for new materials and also reducing carbon emissions in the upstream production process. Second, by using recycled materials and optimizing the production process, the goals of energy and water conservation were achieved. In terms of supply chain management, suppliers replaced plastic pallets with renewable wooden pallets, and encouraged the use of degradable aluminum-plastic bags, environmentally friendly paints, inks, etc., reducing the environmental impact from the source. In terms of logistics optimization, recyclable transfer boxes and new energy vehicles were used for packaging and transportation, reducing carbon emissions during the logistics process. Based on the above self-emission-reduction efforts, the product's carbon neutrality was further achieved by canceling an equivalent amount of Certified Emission Reductions (CERs) under the Verified Carbon Standard (VCS).

Chapter 6
The Application of Key Carbon Emission Reduction Technologies in Carbon Neutrality in Key Areas for Emission

1 Energy Area

1.1 The Characteristics of Emissions and the Application of Key Carbon Emission Reduction Technologies in Electricity Generation

a. Emission Characteristics

The power industry is a cornerstone industry of China, supporting the development of the national economy. Currently, China's electric power generation mainly relies on thermal power generation, which results in a great amount of fossil fuel consumption and contributes the largest share of carbon dioxide emissions within the total carbon emissions in China. Therefore, it is evident that achieving carbon emission reduction in the power industry is crucial for achieving the goals of carbon peaking and carbon neutrality.

As shown in statistics, the national electric power generation exceeded 7.5 trillion kilowatt-hours in 2019, with thermal power accounting for 69.6%, hydropower for 17.4%, nuclear power for 4.6%, wind power for 5.4%, and solar power for 3%. The share of thermal power generation has significantly decreased compared to 2010 (79.2%).

In terms of power supply structure, in 2019, the installed capacity of thermal power accounted for 59.22%, hydropower 17.73%, wind power integrated into the grid 10.45%, solar power integrated to the grid 10.18%, and nuclear power 2.42%. The share of installed thermal power capacity has also significantly decreased compared to 2010 (73.45%). Among all the thermal power installations, coal-fired power generation accounted for about 87.35%, natural gas power generation approximately 7.58%, and biomass power generation around 1.89%.

It is evident that the proportion of non-fossil energy in power generation and installed capacity in China continues to rise and the structure of installed power

C. Chen et al., *Sustainable Carbon Neutrality*,
https://doi.org/10.1007/978-981-95-1691-9_6

generation capacity has also been continuously optimized. However, thermal power generation still accounts for a large proportion, and it is currently the largest carbon emission contributor among all electricity generation approaches. The carbon dioxide emissions per unit electricity generated from thermal power are 2–3 times higher than those from natural gas power, while hydropower, wind power, solar power, nuclear power, and other power sources produce negligible carbon emissions.

b. Typical Emission Sources

The total carbon emissions from thermal power production enterprises include the carbon dioxide emissions from fossil fuel combustion, the desulfurization process for coal-fired power plants, and the use of net purchased electricity of enterprises.

The combustion process of fossil fuels includes emissions from the combustion of coal, oil, and gas consumed by stationary sources (such as boilers and units) and mobile sources. For conventional coal-fired units, it mainly refers to the combustion of conventional coal, such as bituminous coal, lignite coal, and washed coal, in boilers to generate thermal energy. For unconventional coal-fired units, it mainly refers to the process of mixing coal gangue, coal slurry, coal water slurry, and other materials in certain proportions and burning them in boilers to generate thermal energy. For gas-fired units, it mainly involves the process of burning natural gas in gas boilers to generate thermal energy.

The desulfurization process refers to the carbon emission process caused by the consumption of carbonates (such as limestone powder, hydrated lime, soda ash, etc.) in the desulfurizers used in the desulfurization process in boilers of the coal-fired units.

The process of the use of net purchased electricity refers to the carbon emission process from the purchased electricity being used for office operations, daily living, environmental protection facilities, etc. in the state of unit maintenance and shutdown.

Table 1 shows the proportion of carbon emissions from thermal power enterprises. Compared to other industries, carbon emissions from the electric power industry have a single source, i.e. most of the emissions coming from fossil fuel combustion, and only a small part coming from the use of carbonate desulfurizers and the use of net purchased electricity of enterprises.

c. The Application of Key Carbon Emission Reduction Technologies

According to the "2020 Annual Development Report of China's Power Industry", from 2006 to 2019, the total carbon dioxide emissions of China's power industry was effectively reduced, with a 37.0% contribution from reduced coal consumption in electricity generation and a 61.0% contribution from the development of non-fossil energy sources.

Therefore, in order to achieve carbon peaking and carbon neutrality, more efforts should be made to adjust the power supply structure in the field of electricity generation, and continuously develop a modern energy system that is clean, low-carbon, safe, and efficient. At the same time, achieve carbon reduction through technological means such as improving coal-fired power efficiency and energy conservation.

Table 1 Proportion of carbon emissions and their sources of thermal power enterprises

Process	Emission source	Proportion of carbon emissions
Emissions from fossil fuel combustion	Stationary sources: consumption of coal, oil, gas and other stationary sources such as bituminous coal, lignite coal, washed coal, gangue, coal slurry, coal water slurry, diesel, coke oven gas, natural gas, etc. Mobile sources: consumption of mobile sources such as diesel and gasoline	96 ~ 100%
Emissions from desulfurization process	Carbonates such as limestone powder, hydrated lime, soda ash	0 ~ 3%
Emissions from the net purchased electricity	Electricity	0 ~ 1%

(i) Hydroelectric Power Generation

After a long period of development, China's hydropower industry has taken a leading position globally, with its total installed capacity and annual electric power generation consistently ranking first in the world. Remarkable achievements have also been made in terms of hydropower related technologies. To promote further sustainable development of hydropower, leading technologies can be adopted in various aspects to enhance hydropower development capabilities under complex conditions, improve the operational efficiency of hydropower stations, and ensure safe and environmentally friendly operation of hydropower. In terms of project construction, it includes dam construction technologies suitable for cold and high altitude/high seismic magnitude areas with complex geological conditions, seismic resistance and prevention technologies for high dam engineering, construction technologies for super large hydropower projects in cold and high-altitude areas, and building materials for ultra-high dams. In terms of independent manufacturing of water-turbine generator units, it includes large-scale hydroelectric generating units with a capacity of over one million kilowatts, variable-speed pumped storage units, and large-scale impact water-turbine generator units with a capacity of 500,000 kW and a water head of over 1000 m. In terms of ecological protection and restoration technologies, it includes key technologies such as layered water intake, fish passage, habitat construction, artificial breeding and domestication of rare and endemic fish, ecological regulation, vegetation restoration and soil and water conservation in high-altitude areas, etc. Additionally, it covers ecological environment monitoring and control for watershed hydropower development, as well as ecological reconstruction and restoration technology for reservoir drawdown zones and downstream rivers. With regard to "Internet plus" smart hydropower stations, it includes digital watersheds and digital hydropower, "Internet plus" smart hydropower stations and smart watersheds, information management platform technologies, etc. As for the construction of safety monitoring and management systems for the operation of hydropower dams, it involves online safety monitoring and remote technology for dams with a height of over 100 m and a storage capacity of over 100 million cubic meters.

(ii) Solar Power Generation

Solar photovoltaic power generation technology has made breakthroughs in the past decade, with substantial reductions in its manufacturing costs and overall power generation costs, becoming the most cost-competitive energy source in most countries. At this point, China's newly constructed solar PV projects has achieved parity with hydraulic power and thermal power in grid connection, and there is still room for further cost reduction in the future. The future development of solar power generation technology mainly focuses on the following aspects.

a. Technological upgrading and innovation in photovoltaic cells.

At present, the PERC (Passivated Emitter Rear Cell), which is basically a mature technology, gradually replaces BSF (Back Surface Field) cells in the market due to its low cost and high efficiency, becoming the mainstream technology in the market. PERC, also known as Passivated Emitter Rear Cell, utilizes a passivation layer formed by SiN_x or Al_2O_3 on the back of the cell as a back reflector, to enhance the absorption of long wave light and minimize the electric potential difference between P-N electrodes, thus reducing electron recombination and improving the conversion efficiency of solar cells.

Heterojunction with Intrinsic Thin-layer (HIT), whose full name is silicon heterojunction solar cell, is a promising solar cell technology in the industry. It is characterized by the introduction of an intrinsic amorphous silicon layer between the emitter and the substrate and between the highly doped layer on the back and the substrate, which increases the open circuit voltage and improves the conversion efficiency. HIT offers numerous technological advantages, including straightforward process, double-sided power generation, no attenuation, and the possibility of being made thin, which endow it with relatively high development potential. However, it still does not have a cost advantage in large-scale mass production at present. HIT technology is expected to become a leading technological route in the future.

b. The silicon chips are becoming more efficient, thinner, and larger in size, so as to improve power density and conversion efficiency, and achieve cost reduction per kilowatt hour.
c. Develop more utilization scenarios, such as agrivoltaics, aquavoltaics, hydro-PV Hybrid Systems, and wind-Solar Hybrid Systems, etc., to solve land use challenges of photovoltaic power plants, maximize the comprehensive utilization of natural conditions, and achieve intensive development.
d. Promote the zero-pollution development model of the industrial chain: starting from the perspective of whole life cycle, adopting technical measures such as "clean energy for clean energy" and hazard-free recycling in the manufacturing and recycling stages of photovoltaic products, to achieve zero carbon and zero pollution in the entire photovoltaic industrial chain.

(iii) Wind Power

There are abundant wind energy resources in China, with a total reserve of approximately 3.226 billion kilowatts. The wind energy reserves on land that can be developed and utilized are 253 million kilowatts, and the offshore wind energy reserves that can be developed and utilized are 750 million kilowatts, totaling about 1 billion kilowatts.

In recent years, the cumulative installed capacity of wind power generation in China has continued to grow, with a total installed capacity of 210 million kilowatts nationwide. Among them, onshore wind power accounts for the majority with an installed capacity of 204 million kilowatts, accounting for 97%; while offshore wind power has an installed capacity of 5.93 million kilowatts, accounting for 3%. Wind power generation is also an important means of utilizing renewable energy in the future, with its technological development direction mainly focusing on the following four aspects.

a. Developing large-scale wind turbines. If the wind resources are sufficient, increasing the capacity of individual wind turbines can effectively reduce infrastructure investment, increase the capacity coefficient of wind turbines, and reduce the costs of power generation per unit. At present, some manufacturers have launched wind turbines with a single-unit capacity of 15 trillion watts, with a wind wheel diameter of 236 m and a power generation capacity of up to 80 GWh a year. The capacity factor is expected to exceed 60%.
b. Pushing the development of offshore wind power. Currently, the economically exploitable wind resources on land are diminishing, while the offshore wind energy reserves in China that can be developed and utilized are three times greater than those on land. Additionally, the energy efficiency of offshore wind energy resources is 20 ~ 40% higher than those of onshore wind farms. Besides, they do not occupy land resources and have high wind speed, making them suitable for large-scale development. The R&D directions for offshore wind power technology mainly include ultra-long and ultra-flexible blade technology, main bearing technology, hydraulic pitch control technology, supporting structure technology, flexible DC transmission and transformation integration technology, offshore wind farm cluster control technology, and intelligent operation and maintenance technology for offshore wind power.
c. Application of smart wind power technology. The collection and aggregation of real time wind farm operation data are realized through digital technology. Based on data analysis, the turbine load and power generation output can be effectively balanced, thus maximizing the performance of the units. An intelligent monitoring platform enables the assessment of the functioning state of the units, allowing for precise detection of operational conditions, which enhances the accuracy of fault prediction and improves the reliability of the units. Technologies such as big data and machine learning are utilized to facilitate the post-operation evaluation of wind farms, thus guiding rapid iterations of unit design and rational layout of wind farms.

d. Research and development of high-altitude wind power generation technology. High-altitude wind power generation technology is considered to be a key direction for the future development of wind power, and may become the world's largest source of electricity. Compared to the quality of traditional onshore and offshore wind resources, the advantages of high-altitude wind energy include stable airflow, high wind speed, abundant reserves. At the same time, high-altitude wind power does not require large supporting structures and occupies only a small area. At present, high-altitude wind power generation is still in the stage of technological research and development, and the commercialization progress is relatively slow. However, its future development prospects remain promising.

(iv) Nuclear Power

The fusion of light atomic nuclei and the fission of heavy atomic nuclei can both release energy, known as nuclear fusion energy and nuclear fission energy respectively. During fusion or fission, a large amount of heat is generated, and the energy is converted from nuclear energy to mechanical energy and then to electric energy. This type of electricity is called nuclear power.

Nuclear fission power generation technology has been successfully commercialized for decades, with nuclear power accounting for about 10% of global electricity generation. Nuclear power is a typical carbon-free energy source that generates electricity without CO_2 emissions. In France, the proportion of nuclear power in its power supply is the highest in the world, with nuclear power accounting for more than 70% of its total electricity generation, making France one of the countries with the lowest carbon intensity in global power industry. However, concerns related to the safety of nuclear fission and the challenges of nuclear wastes disposal have hindered its widespread application. At present, the proportion of nuclear power generation in China is less than 5%. Therefore, developing nuclear power proactively and orderly while ensuring safety, is one of the important ways for China to achieve the goal of carbon neutrality.

Controllable nuclear fusion technology is regarded as the ultimate solution to address energy problems. On Earth, deuterium, the raw material for nuclear fusion, exists abundantly in seawater, essentially making it an inexhaustible resource. Moreover, the nuclear fusion reaction process does not produce pollutants or hazardous wastes, nor does it emit greenhouse gases. Meanwhile, the safety level of nuclear fusion reactors is remarkably high. Consequently, owing to its many merits such as plenty, safety, environmental friendliness and high efficiency, controllable nuclear fusion energy basically meets the various requirements of humanity for the ideal ultimate energy source in the future. However, currently, controllable nuclear fusion technology is still in the research stage, and long-term efforts of scientists from various countries are needed to achieve technological breakthroughs.

(v) Phasing Out Coal-Fired Power Generation and Encouraging the Use of Gas-Fired Generation Units

Among the three major fossil fuels, namely coal, oil and natural gas, natural gas stands out with the highest hydrogen content, high thermal energy utilization efficiency and the highest calorific value per unit mass. Its carbon emissions are only half of those of coal. According to the "Implementation Plan for the Total Quantity Setting and Allocation of National Carbon Emission Rights Trading Quotas in 2021 and 2022 (Power Generation Industry)", in 2022, for gas-fired unit sets, the benchmark value for the power supply intensity was 0.3901 tCO_2/MWh, and the benchmark value for heat supply intensity was 0.0557 tCO_2/GJ. In comparison, for conventional coal-fired unit sets, the benchmark value for power supply intensity ranged from 0.8177 to 0.8729 tCO_2/MWh, and the benchmark value for heat supply intensity was 0.1105 tCO_2/GJ. To promote the development of gas-fired unit sets, in the allowance settlement process of gas-fired unit sets, if the verified emissions of gas-fired unit sets are no less than the allocated free allowance, their allowance settlement obligation is the entirety of the free allowances they have received; if the verified emissions of gas-fired unit sets are lower than the allocated free allowance amounts, their allowance settlement obligation is equivalent to the amount of verified emissions of the gas-fired unit sets. In short, under the current national carbon emission trading system, gas-fired unit sets seem to enjoy all the advantages without incurring any real costs.

At the current stage, phasing out coal-fired power generation units in the electric power industry, especially those with poor energy efficiency, and replacing them with gas-fired power generation unit is one of the crucial pathways to peak carbon emissions in the power sector as soon as possible.

(vi) Ultra-supercritical Coal-Fired Power Generation Technology

Ultra-supercritical coal-fired power generation technology is an advanced and highly efficient power generation technology, with a thermal efficiency approximately 4% higher than that of supercritical units. It demonstrates more prominent advantages over conventional coal-fired power generation units. The working fluid within the boiler is water, and the critical parameters of water are 22.129 MPa and 374.15 °C. At this pressure and temperature, the density of water and steam is identical, which is known as the critical point of water. When the boiler steam temperature is not lower than 593 °C or the steam pressure is not less than 31 MPa, it is called as ultra-supercritical. The ultra-supercritical units have the following advantages:

(a) High thermal efficiency. The net efficiency of ultra-supercritical units can reach around 45%, which can save energy, cut down energy consumption, and reduce carbon emissions.
(b) Reduced pollutant emissions. Due to the measures like desulfurization, denitrification, low-nitrogen combustion, and the installation of highly efficient dust collectors, the concentration of pollutant emissions is significantly decreased, thus meeting the ultra-low emission standards.
(c) Large single-unit capacity. The capacity of ultra-supercritical units is generally at the level of one million kilowatts.

(vii) Double Reheat Technology of Coal-Fired Power Generation Units

The double reheat technology is widely recognized as an effective method to improve the efficiency of coal-fired power units. Compared to single reheat, double reheat introduces an additional reheat process based on the single reheat process, which increases the average endothermic temperature within the power generation cycle, thus boosting the power generation efficiency. With the improvement of power generation efficiency, the carbon dioxide emissions per unit of electricity generated are significantly reduced. Under the same steam pressure and temperature, the thermal efficiency of double reheat units is 2% higher than that of single reheat units, resulting in a corresponding carbon dioxide reduction of approximately 3.6%. Therefore, the double reheat technology is a feasible thermal power generation technology for energy conservation, emission reduction, and environmental protection.

1.2 The Emission Characteristics and the Application of Key Carbon Reduction Technologies in Power Transmission

a. Emission Characteristics

China has built the world's largest interconnected power grid, with an ultra-high voltage power grid as its backbone network. Through the coordinated development of power grids at different levels, it has achieved the transmission of electricity from the west to the east, from the north to the south, the mutual supplementation of hydropower and thermal power, and the complementary utilization of wind and solar power. It boasts world-leading energy allocation capabilities and a high level of safe power supply through the grid, effectively supporting the sustainable development of China's economy and society while ensuring the electricity demands of the general public are met.

However, the distribution of China's energy resources and load centers is highly unbalanced. Generally speaking, there are more energy resources in the west and the north while fewer in the east and the south. However, the demand for electricity has been predominantly concentrated in the central and eastern regions for a long time. More than 80% of China's energy resources are located in the western and northern regions, while 75% of the power consumption is concentrated in the eastern and central regions. As a result, the amount of long-distance cross-region power transmission in China has been continuously increasing, with high-capacity, long-distance transmission being a key feature of our power grid structure.

Due to the large territory, unbalanced distribution of resources and loads, and long transmission distances, the total power loss in China's power grid is significant. At the same time, as the electricity currently entering the grid is still mainly generated by coal-fired power plants, the carbon dioxide emissions generated during the power generation process attributed to the power grid losses are also huge. In addition, the use of related electrical equipment for long-distance transmission also leads to the widespread use of greenhouse gases such as sulfur hexafluoride (SF_6).

b. Typical Emission Sources

The greenhouse gas emissions of power transmission enterprises mainly include the emissions of sulfur hexafluoride (SF_6) from the repair and decommissioning of equipment using SF_6, as well as the carbon dioxide emissions from power generation process associated with power transmission and distribution losses.

Sulfur hexafluoride (SF_6) is widely used in the electrical equipment related to power transmission, such as circuit breakers, high-voltage switches, high-voltage transformers, gas-insulated switchgear, high-voltage transmission lines, and current transformers, due to its excellent insulating and arc-extinguishing properties. Although the amount of SF_6 gas contained in electrical equipment is relatively low, its Global Warming Potential (GWP) is 23,500 times[1] of that of carbon dioxide, making its contribution to the greenhouse effect significant. Therefore, it is a key emission source that requires stringent control measures.

The carbon dioxide emissions of power grid enterprises mainly come from the greenhouse gas emissions generated by the power losses on power transmission and distribution lines. This loss is calculated based on the difference between the power supplied and the power sold.

c. Application of Key Carbon Emission Reduction Technologies

The primary methods for reducing greenhouse gas emissions, based on the characteristics of emissions from electricity transmission and the types of typical emission sources, include ultra-high voltage transmission, technology for integrating renewable energy power into the grid, and the reduction of sulfur hexafluoride emissions.

(i) Ultra-High Voltage Transmission

According to the Joule's law, the power loss on the transmission line is proportional to the square of the current. By lowering the current on the transmission line, the power loss can be reduced. At the same time, for long-distance transmission, it is necessary to reduce the transmission current to reduce the heat generated by the conductors. However, long-distance transmission power must be high enough to be practically useful. To achieve a reduction in transmission current while maintaining a constant power, the voltage needs to be raised. Therefore, ultra-high voltage transmission offers significant technical advantages, including large transmission capacity, long transmission distances, high efficiency, and low power losses.

Ultra-high voltage (UHV) refers to transmission technologies with a voltage level of 1000 kV and above for alternating current, and $\pm$ 800 kV and above for direct current. Compared to traditional transmission technologies, UHV transmission can increase transmission capacity by up to three times, extend transmission distances by up to 2.5 times, and reduce transmission losses by 45%. Additionally, the corridor width per unit capacity can be decreased by 30%, and the cost per unit capacity can be cut down by 28%. This technology enables a safer, more efficient, and more environmentally friendly energy distribution. According to the estimation of the State Grid Corporation of China, when transmitting the same amount of electricity, using 1000 kV lines can save 60% of land resources compared to using 500 kV lines.

China's UHV technology is at the forefront globally. As of 2020, there were 25 ultra-high voltage lines in operation and 7 under construction in the country, with plans to initiate several more projects in the future. The construction and operation of these ultra-high voltage lines will significantly enhance the transmission efficiency of China's power grid, reduce transmission losses, and cut down carbon dioxide emissions.

(ii) Technology for Integrating Renewable Energy Power into the Grid

Currently, the share of electricity generated from non-fossil energy in China is approximately 30%. According to relevant studies and calculations, to achieve the carbon neutrality goal by 2060, the share of non-fossil energy in primary energy consumption needs to exceed 80%. The share of renewable energy, represented by wind and solar power, is expected to rise to over 50%. However, due to the inherent randomness, intermittency, and volatility of renewable energy power, large-scale integration into the grid will pose significant challenges to the safe and stable operation of the power grid and the balance of electricity supply and demand. It is essential to continuously enhance the balancing and regulating capabilities of the power system, and facilitate the full consumption of new energy sources. Addressing the grid connection issues for renewable energy power and reducing the wind and solar power that could not be integrated into the gird poses substantial challenges to the power transmission system in achieving carbon neutrality.

Through the R&D and implementation of smart grid and energy Internet technologies, which include the adoption of advanced sensing technologies, information and communication technologies, and automated control systems, a new type of modernized grid characterized by high level of information integration, automation, and interactivity is developed. It enables optimized control in various aspects such as renewable energy power generation management, short-term power forecasting, power distribution for systems with a high proportion of renewable energy, and voltage and frequency control of renewable energy power systems. Consequently, this enhances the grid's capacity to accommodate renewable energy, which ensures safe, reliable, economical, and efficient operation of the grid.

By adopting advanced energy storage technologies, the continuity and reliability of power supply can be ensured through timely energy storage and release. Pumped storage hydropower plants utilize surplus electricity during low demand periods to pump water to an upstream reservoir, which is then released to a downstream power plant during peak demand periods to generate electricity. This process transforms excess electricity during low load periods into high-value electric energy during peak periods. Additionally, these plants offer functionalities such as peak shaving, frequency regulation, phase adjustment, energy storage, system backup, and black start, along with advantages like large capacity, system friendliness, economy, reliability, and ecological protection. Pumped storage is currently the most mature large-scale energy storage method and is a crucial component of the new power system dominated by renewable energy. China leads the world in installed capacity for pumped storage hydropower, yet its proportion remains relatively low. Moving forward, the development of pumped storage will continue to accelerate, addressing the challenges of renewable energy consumption, integration into the grid and peak regulation. Besides pumped storage, physical energy storage technologies such as compressed air energy storage, flywheel energy storage and superconducting energy storage, chemical reaction energy storage technologies such as hydrogen storage, carbon storage and synthetic fuel storage, and electrochemical energy storage technologies such as energy storage batteries and supercapacitors, along with heat and cold storage technologies such as sensible heat storage, phase change storage and ice storage for cooling are also expected to support the rapid development of renewable energy generation and utilization.

(iii) Reduction of Sulfur Hexafluoride Emissions

SF_6 (Sulfur hexafluoride) emission reduction technologies mainly include emission reduction and equipment replacement.

Firstly, reducing the emission of SF_6 gas in the production, maintenance, and decommissioning process of equipment. During the generation of SF_6 gas, emissions of SF_6 can be reduced at the source by process optimization and improvement. During the manufacturing of SF_6-using electrical equipment, measures such as process control of gas transmission and filling, along with leakage inspections, should be implemented to minimize gas loss. At the same time, enhancing manufacturing process and performing strict quality inspection before delivery can reduce the leakage and dispersion of the gas. During the operation of SF_6-using electrical equipment, carrying out management at a detailed level and strengthening equipment status monitoring and maintenance can help extend the service life of the equipment. For decommissioned SF_6-using electrical equipment, effective technical means should be adopted to recover, purify, and recycle the gas, thus preventing any emissions.

Secondly, the search for alternative gases to SF_6 is one of the hot research directions in the power transmission industry. Currently, the main focus is on alternative gases such as conventional gases (air, N_2, and CO_2), SF_6 mixtures, and highly electronegative gases along with their mixtures. Some alternative solutions have been implemented in medium and low-voltage equipment; however, the technology for completely replacing SF_6 is not yet mature and requires further research and exploration.

1.3 The Characteristics of Emissions and the Application of Key Carbon Reduction Technologies in Coal Production

a. Emission Characteristics

Coal is the primary energy source in our country, accounting for over 50% in the primary energy structure. China is also the world's largest producer, consumer, and importer of coal. In addition to carbon emissions generated during coal combustion, a large amount of methane (CH_4) will also leak during the coal mining process.

Given that methane has a Global Warming Potential (GWP) 28 times[1] that of carbon dioxide, it contributes substantially to the greenhouse effect.

The amount of methane leakage from coal mines depends on multiple factors such as mine depth, coal quality, and the geological age of coal seams. The deeper the mine and the older the coal seam is, the higher the methane concentration is, leading to more emissions and leakage during the mining process. In China, the large number of deep coal mines contributes to relatively high methane emissions. Additionally, due to the substantial coal production, China's total methane emissions from the coal industry rank first in the world.

Compared to methane leakage in the oil and gas sector, managing methane leakage in coal mining presents greater challenges. In the oil and gas industry, the economic benefits of methane recovery are significant, with 45% of methane leakage achieving net-zero cost emission reduction. However, in the coal sector, the concentration of leaked methane is lower and fluctuates a lot, making the recovery and utilization more technically complex and costly. All these factors make methane emission reduction in coal mining technically and economically challenging.

b. Typical Emission Sources

The main emissions from coal production include emissions from fuel combustion, flaring emissions, fugitive emissions, underground mining emissions, open-pit mining emissions, emissions from post-mining activities, and the implicit emissions from net purchased electricity and heat.

Fuel combustion emissions refer to emissions generated in the burning of fossil fuels in various types of stationary or mobile combustion equipment.

Flaring emissions refer to the greenhouse gas emissions resulting from the flare treatment of coal mine gas (coalbed methane) from coal mining before it is emitted for safety, environmental protection purposes.

Fugitive emissions refer to the intentional or unintentional discharge of CH_4 and CO_2 from coal during coal mining, processing, and transportation, including emissions from underground mining, open-pit mining, and post-mining activities.

Emissions from underground mining refer to the emissions of CH_4 and CO_2 that are stored in coal seams. These gases continuously enter into the mine tunnels and mining spaces during underground mining operations and are then emitted into the atmosphere through ventilation and drainage systems.

Emissions from open-pit mining refer to the CH_4 emissions released during open-pit mining and released from adjacent exposed coal (soil) layers.

Emissions from post-mining activities refer to the CH_4 emissions generated by the slow release of residual gas in coal during coal washing, selection, storage, transportation, and crushing before combustion.

Implicit emissions from net purchased electricity and heat refer to the CO_2 emissions generated by fuel combustion during the production process of the net purchased electricity or heat (steam, hot water).

c. The Application of Key Carbon Reduction Technologies

Based on the characteristics of coal production, the key emission reduction technology is to reduce methane emissions.

(i) Coalbed Methane (Coal Mine Gas) Mining

During the "13th Five-Year Plan" period, significant advancements have been made in the management and mining technologies of coalbed methane (coal mine gas) in China. The control of coal mine gas has evolved from local to reginal, from only underground mining to both underground and open-pit operations. This has gradually led to the development of advanced gas management models for typical geological conditions, such as the co-extraction of coal and gas at the protective layers and the interactive three-zone (planning zone, preparation zone, and production zone) gas extraction above and under the ground. Technological innovations have been widely applied and promoted throughout the industry, such as efficient drilling, hydraulic permeability enhancement, and tiered gas utilization. New breakthroughs have been achieved in the evaluation of medium and low-rank coalbed methane resources, the extraction of deep coalbed methane, and the co-extraction of coalbed methane, gas in tight sand, and shale gas. Significant progress has been made in technological innovations such as cross layer fracturing of vertical wells, staged fracturing horizontal well, and intelligent drainage and extraction, resulting in a substantial increase in single well output.

(ii) Utilization of Coalbed Methane (Coal Mine Gas)

The recovery and utilization of coalbed methane (coal mine gas) is an important means to reduce methane emissions during the coal mining process. According to relevant national standards, emissions of high-concentration gas (methane volume fraction $\geq 30\%$) are prohibited. It is essential to establish gas storage tanks and build supporting facilities for gas utilization. For gas that cannot be utilized, it should also be treated through flaring burning, and direct emission is prohibited.

Currently, the recovery and utilization of high-concentration coalbed methane (coal mine gas) have been widely implemented. It is used as an industrial fuel and a chemical raw material, or used in fields such as power generation and civil use, becoming a clean and efficient new energy. However, the purification or utilization techniques for low-concentration gas are still immature, and have not yet been widely commercialized due to their high technical costs. This represents a key area for future technological breakthroughs.

1.4 The Characteristics of Emissions and the Application of Key Carbon Reduction Technologies in Oil and Gas Extraction

a. Emission Characteristics

The oil and gas industry serves as both an energy producer and consumer, 25 generating carbon emissions throughout the processes of extraction, transportation, storage, and end-use applications. In 2018, the total energy consumption of the oil and gas extraction sector reached 38.18 million tonnes of standard coal,[1] accounting for 1.23% of the total energy consumption in the national industrial sectors, resulting in approximately 94.3 million tonnes of CO_2 emissions.

Carbon emissions from oil and gas extraction primarily consist of two parts: CO_2 and CH_4. The CO_2 emissions mainly come from the heating and energy supply demands during the production process, while the CH_4 emissions mainly come from gas leakage during the extraction and transportation of oil and gas. Although the volume of CH_4 emissions is significantly lower than that of CO_2, its Global Warming Potential is 25 times[1] that of carbon dioxide. The categories of emission sources and types of gas emissions for oil and gas extraction companies include the following[2]:

(i) CO_2 Emissions from Fuel Combustion

CO_2 emissions from fuel combustion mainly refer to the emissions generated in the combustion process of fossil fuels used for power or thermal supply in the links of oil and natural gas exploration. For example, emissions generated by the self-built power stations and heating stations on offshore platforms when using associated gas, crude oil, and other fuels to meet their own electricity and heating demands.

(ii) Emissions from Flaring

In oil and gas extraction enterprises, various production activities generate combustible waste gases, including those generated during working conditions such as normal operations, maintenance, well testing, and accidents. If these gases cannot be recovered or transported out after being collected, they are typically directed to one or more flare systems for combustion treatment prior to discharging, for the sake of environmental protection and safety. Flaring may result in CO_2 emissions, along with the possibility of minor CH_4 emissions.

(iii) Emissions from Process Venting and Discharging

Process venting and discharging emissions mainly refer to the intentional release of CH_4 or CO_2 gas into the atmosphere through the release ports or safety valves on the process devices in various links of oil and gas extraction, such as natural gas emissions that drive the operation of pneumatic devices, pressure relief emissions, equipment blowdown emissions, process exhaust emissions, and tank dissolved gas emissions.

(iv) CH_4 Fugitive Emissions

CH_4 fugitive emissions mainly refer to the uncontrolled CH_4 emissions caused by equipment leakage in various links of oil and gas extraction, such as leakages from valves, flanges, pump impeller seals, compressor seals, pressure reducing valves, sampling interfaces, process drainage, open pipes, casings, storage tanks, and other pressure equipment leakages that are not defined as process venting.

(v) Amount of CH_4 Recovered and Utilized

The amount of CH_4 recovered and utilized mainly refers to the part of CH_4 that enterprises recover from the waste gas flows vented and discharged during the production process through energy-saving and emission-reduction technologies, which prevents this part of CH_4 from being emitted into the atmosphere. The amount of CH_4 recovered and utilized can be deducted from the total emissions of enterprises.

(vi) Amount of CO_2 Recovered and Utilized

The amount of CO_2 recovered and utilized mainly refers to the part of CO_2 that enterprises recovered from fuel combustion or process venting and discharging, which is used as raw materials for production or products supplied to the market, which prevents the emission of CO_2 into the atmosphere. The amount of CO_2 recovered and utilized can be deducted from the total emissions of enterprises.

(vii) Implicit CO_2 Emissions from Net Purchased Electricity and Heat

This part of the emissions actually occur in the enterprises that produce these electricity or heat, but arise from the consumption activities of the greenhouse gas emission reporting entities. Currently, according to the agreement, these emissions are accounted under the reporting entity's name when calculating carbon emissions.

b. Application of Key Carbon Reduction Technologies

Based on the characteristics of carbon emissions from oil and gas extraction, carbon reduction in this regard can be achieved through optimizing energy structure, adopting advanced oil and gas extraction technologies, improving emission processes, recovering and utilizing waste heat, and applying comprehensive CO_2 utilization technologies. The key lies in reducing emissions from the combustion of fossil fuels, as well as the venting and leakage of natural gas.

(i) Optimizing Energy Structure

During the oil and gas extraction process, most of the energy is consumed in the electricity generation and heat supply process related to production and daily life. Replacing crude oil and diesel with cleaner natural gas is an effective means of reducing greenhouse gas emissions. Renewable energy sources can also be developed and utilized to replace fossil energy, such as wind energy, hydro energy, biomass energy, ocean energy, tidal energy, and geothermal energy, as well as other new types of green and low-carbon energy sources.

Electricity is one of the main energies in oil and gas extraction process. Oil and gas companies have built a large number of independent power plants to meet their own needs, resulting in a low reliance on grid-based electricity. Compared to self-generated electricity, grid-based electricity typically has a lower greenhouse gas emission factor. Oil and gas field enterprises are often located in areas with high-quality resources, such as wind power, photovoltaic power, hydroelectric power, etc. Promoting the use of grid-based electricity for drilling and fracturing is the main direction for oil and gas field enterprises to save energy, reduce emissions and costs, enhance efficiency and achieve clean development.

(ii) Adoption of Advanced Oil and Gas Extraction Technologies

The use of Enhanced Oil Recovery (EOR) technology can extract 30–60% or more of the oil originally stored in the oil pool. Common EOR methods include thermal oil recovery, chemical oil displacement, gas solvent oil displacement, and microorganism oil recovery. By enhancing oil recovery, an additional 5–20% of oil can be recovered.

Solar EOR technology is a thermal oil recovery method that concentrates sunlight on heat pipes to produce the steam needed for oil production. In the solar EOR market, GlassPoint is undoubtedly the industry leader. Its unique closed parabolic trough heat collector design greatly reduces investment and operating costs. This approach is more cost-effective than generating steam with natural gas, allowing oil fields to achieve up to an 80% reduction in natural gas consumption. For the recovery of many heavy oil fields around the world, fuel costs can account for 60–80% of the operational expenses. Using solar energy to produce steam is cheaper than burning fossil fuels, and compared to traditional EOR technologies, it reduces the generation of pollutants such as nitrogen oxides, carbon dioxide, and particles.[1]

Carbon dioxide oil displacement is a gas solvent oil displacement method that represents a mature technology for enhancing oil recovery at the later stages of oil pool exploitation. Liquid CO_2 serves as a unique oil recovery agent due to its high solubility in crude oil and its supercritical state within the stratum, making it easy to form a miscible phase with the crude oil. This process increases the capillary number and reduces interfacial tension, while the introduction of gas also decreases the viscosity of the crude oil, significantly enhancing oil recovery.[1] Injecting CO_2 as the main gas into the oil reservoir for oil displacement and storing it in the reservoir not only effectively mitigates the greenhouse gas effect, but also effectively enhances oil recovery.

(iii) Improving Releasing Process

Associated gas typically refers to the natural gas that coexists with oil. In the context of oil extraction operations, it refers to the natural gas produced during the extraction of oil fields or oil pools. Associated gas is a significant source of flare gas; however, due to various factors such as the lack of systems for the collection, storage, and export of associated gas, as well as space and safety constraints on offshore platforms, associated gas is generally discharged through flaring. To enhance the recovery and utilization of associated gas resources, measures such as gas recovery and export, recovery for power generation and heating, recovery for reinjection, and the production of CNG or LNG can be implemented.

When hydrocarbon gases cannot be recovered, it is advisable to avoid venting them directly, but instead, discharge them using flare systems. The vented gases primarily contain unburned CH_4, which has a Global Warming Potential 25 times that of CO_2. Employing flare systems to combust the excess CH_4 generated from oil and gas production systems can generate CO_2 instead of releasing CH_4 directly into the atmosphere, thereby mitigating the greenhouse gas effect.

Controlling CH_4 emissions from the oil and gas systems. Adopting technologies such as green well completion and low-emission gas well drainage; implementing pipeline directional maintenance and pre-maintenance evacuation technology, storage and transportation facility maintenance technology, and light hydrocarbon steam evacuation technology; using high-efficiency compressors and low-emission/zero-emission pneumatic equipment. Promoting uncontrolled emission monitoring of equipment components and adopting advanced sealing technology to reduce uncontrolled emissions from gas well production, pipeline and storage operations, and equipment operations. Promoting casing gas/associated gas recovery technology and the light hydrocarbon steam recovery technology for crude oil storage tanks. Carrying out comprehensive CH_4 leakage monitoring and CH_4 recovery and utilization operations that covers oil and gas production, processing, and storage and transportation businesses, so as to improve the CH_4 recovery and utilization rate of the oil and gas systems.

(iv) Recovery and Utilization of Waste Heat

Abundant waste heat resources are typically found in offshore gas fields, mainly coming from two primary sources. First, the high-temperature waste heat from gas turbines that drive the generator sets; second, the high-temperature waste heat from gas turbines that power the natural gas compression units. Calculations indicate that the efficiency of gas turbines is approximately 30%. Around 30% of the heat is lost through the heat dissipation of the units, and another 30% of energy is discharged along with the high-temperature exhaust gases, remaining unutilized. Particularly in the later stages of gas field extraction, as reservoir pressure declines, in order to ensure the normal transportation through the offshore natural gas processing facilities and submarine natural gas pipelines, natural gas compressors driven by high-power gas turbines are needed. As a result, the waste heat resources are considerable.[1]

(v) Technologies for Comprehensive utilization of CO_2

The oil extraction industry has significant potential in the comprehensive utilization of CO_2. Techniques such as CO_2 oil displacement, CO_2 fracturing, carbon capture and storage (CCS), and CO_2 reinjection fall under this category.

CO_2 oil displacement involves using captured CO_2 to facilitate oil displacement. Integrating CCS projects with CO_2 EOR projects can not only increase the oil recovery but also achieve carbon storage, thereby fulfilling the dual objectives of reducing carbon emissions and increasing oil and gas production.

The carbon dioxide replacement method refers to the injection of CO_2 into the natural gas hydrate reservoir by taking advantage of the property that CO_2 is more likely to form hydrates, and then the natural gas hydrates are decomposed to form

CO_2 hydrates. This method can not only extract natural gas, but also sequester the greenhouse gas CO_2 and greatly reduce the possibility of geological disasters caused by hydrate decomposition.

Carbon dioxide fracture flooding is used to extract unconventional oil and gas resources. This technology can not only conserve water, reduce fracturing pressure, prevent reservoir damage, and enhance oil and gas recovery rates, but also facilitate geological storage of part of CO_2, thus achieving green and low-carbon development of unconventional oil and gas resources.

2 Industrial Area

2.1 *The Characteristics of Emissions and the Application of Key Carbon Reduction Technologies in the Steel Industry*

a. Emissions Characteristics

China is the largest steel producer in the world, maintaining the top position in crude steel output for over twenty consecutive years. Additionally, the steel industry is a major consumer of energy, accounting for approximately 15% of the total industrial energy consumption in the country.

In 2019, global carbon emissions from the steel industry reached around 2.6 billion tonnes per year, representing about 7% of total emissions from the global energy system. China's steel industry alone emitted 1.8 billion tonnes of carbon, which constitutes over 60% of the global steel industry's emissions and approximately 15% of the nation's total carbon emissions, making it the highest among the 31 manufacturing sectors. Over the past twenty years, the steel sector has made advancements in technologies, processes, and equipment, as well as in energy conservation, environmental protection, and intelligent operations. In 2019, the energy consumption per tonne of steel in the industry was 553 kg of standard coal per tonne, showing a significant improvement in energy efficiency compared to 750 kg of standard coal per tonne a decade earlier.

b. Typical Emission Sources

(i) Production Process

The production processes of a large steel joint production enterprise generally include the following.

(a) Coking: A certain proportion of coal is converted into coke, coke gas, and chemical products through high-temperature dry distillation in the coke ovens.
(b) Sintering: Iron ore powder, coal powder, coke powder, and lime are mixed in a certain proportion and burned to form sintered ore with sufficient strength and grain size.

(c) Pelletizing: A mixture of concentrated ore powder and melting agent (sometimes with binders and fuels) is rolled into green pellets in a pelletizing machine, which are then dried, roasted, and consolidated to form ferrous materials with good metallurgical properties.
(d) Ironmaking: Iron oxides are reduced to metallic iron using a reducing agent.
(e) Steelmaking: Excess carbon and impurities such as sulfur and phosphorus in feedstocks (iron water and scrap steel, etc.) are removed, then appropriate alloying elements are added to the feedstock (including converter steelmaking and electrosmelting of steelmaking, etc.).
(f) Steel rolling and processing (rolling): Steel ingots are put through pressure processing to induce plastic deformation, forming steel products with specific shapes and dimensions.

The steel production process flow is shown in Fig. 1.

(ii) Main Emission Equipment and Facilities

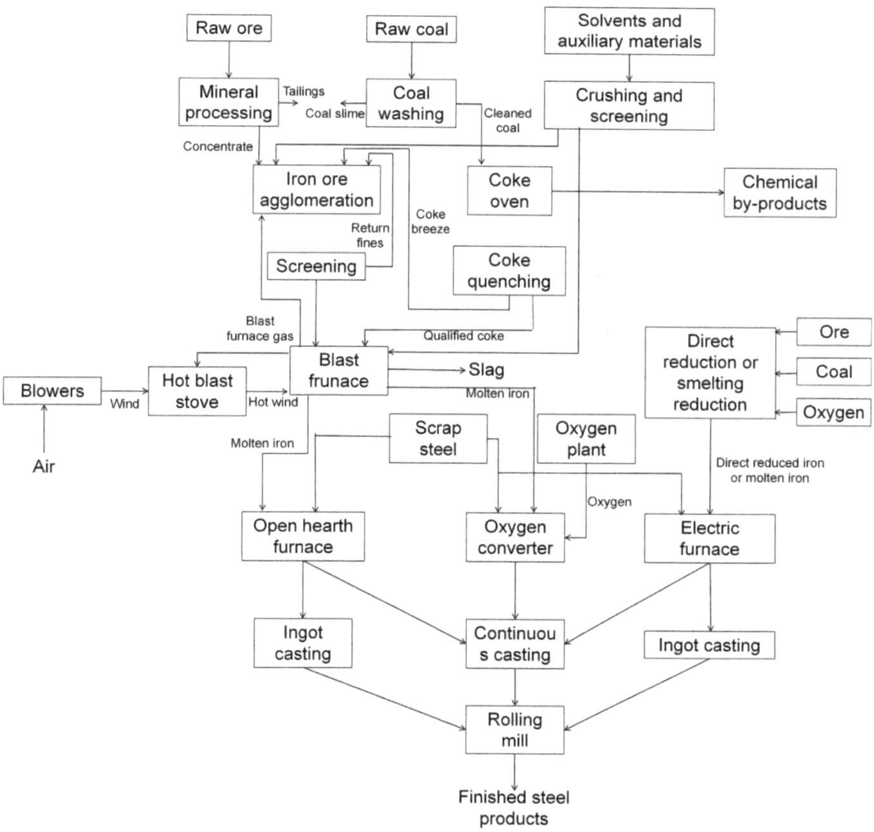

Fig. 1 Steel production process flow

The main emission equipment and facilities for each process of steel production are shown in Table 2.

(iii) Typical Emission Sources in the Steel Industry

(i) Fuel combustion emissions: CO_2 emissions generated from the consumption of fossil fuels, such as the diesel and gasoline fuels consumed by combustion equipment like coke ovens, sintering machines, blast furnaces, and industrial boilers, as well as transportation vehicles and in-house handling equipment used in production.

(ii) Production process emissions: CO_2 emissions generated from the decomposition and oxidation of other externally sourced carbon-containing materials (such as electrodes, pig iron, ferroalloys, and direct reduced iron) and melting agents during processes like sintering, ironmaking, and steelmaking.

Table 2 Main emission equipment and facilities for each process of steel production

Production process	Production equipment (CO_2 emission equipment)
Coking	Equipment and facilities such as crushers, batching, screening, coke oven bodies, coal preheating, coke quenching, coal gas treatment, and recovery workshops (condensation blowing, ammonia recovery, crude benzene recovery, etc.)
Pellets	Equipment and facilities for each production process, from batching, feedstock supply, pelletizing, roasting, screening, to the process in which the finished pellet bell conveyor leaving the pelleting process
Sintering	Equipment and facilities for each production process, from solvent and fuel crushing, through batching, feedstock transportation, process feedstock mixing, sintering machine, sintered ore crushing, screening, and until the finished sintered ore belt conveyor leaves the sintering process
Ironmaking	Equipment and facilities for systems such as feedstock supply, hot-blast stoves, coal powder drying and injection, blast furnace bodies, cast iron, coal gas treatment, slag iron and slag treatment, etc.
Steelmaking	Equipment and facilities for systems such as molten iron pre-treatment, converter bodies, continuous casting (or casting), slag treatment, steel ladle baking, external refining, coal gas treatment
Steel rolling and processing	Equipment and facilities such as soaking pit, heating furnaces, plastic forming equipment (including rolling, forging, extrusion, drawing, and stamping equipment), finishing equipment, welding processing, coating processing, etc.
Self-supplied power plant	Equipment and facilities for crushing, coal grinding, preheating, boilers, steam turbines, generators, etc.
Lime firing	Equipment and facilities for coal grinding, crushing, stone washing, screening, preheating, lime furnaces, pellet making (pellet pressing), etc.
Auxiliary production system	Equipment and facilities for production management and scheduling systems, mechanical maintenance, laboratory testing, metering, various fans, water circulation system, environmental protection (water treatment and dust removal, etc.), oxygen stations, surplus energy power plants, heating boilers and etc.

(iii) Emissions from net purchased electricity and heat: CO_2 emissions implicitly generated from net purchased electricity and net purchased heat (such as steam).
(iv) Implicit emissions from carbon-sequestered products: A small portion of carbon is sequestered in the export products produced by the company during the steel production process, such as pig iron and crude steel. Additionally, a small amount of carbon is sequestered in carbon-sequestered products like methanol, which is produced using by-product coal gas as a feedstock. The CO_2 emissions corresponding to the carbon sequestered in these products should be deducted.

(iv) Proportion of Emissions from Main Processes

The carbon emissions of each process in large-scale steel joint production enterprises are shown in Table 3.

The coking process primarily consumes cleaned coal and produce coke and coke oven gas for the use in other processes. Many steel enterprises do not have a coking process and purchase coke from independent coking plants. For those steel enterprises with a coking process, after deducting the output of coke, coke oven gas, and other by-products, the carbon emissions account for approximately 5%.

The sintering process in steel production enterprises accounts for a large portion of energy consumption and emissions, approximately 10%. In contrast, the pelletizing

Table 3 The carbon emissions of each process in large-scale steel joint production enterprises

Process	Energy consumption (+)	Energy recovery (−)	Proportion of carbon emissions
Coking	Cleaned coal, blast furnace gas, electricity	Coke, coke oven gas, tar, crude benzene	About 5%
Sintering	Cleaned coal, coke, coke oven gas, blast furnace gas, electricity	–	About 10%
Pelletizing	Blast furnace gas, coke oven gas		About 1%
Ironmaking	Anthracite, bituminous coal, coke, coke oven gas, converter gas, electricity	Blast furnace gas	About 25%
Steelmaking (converter)	Coke oven gas, blast furnace gas	Converter gas	About 7%
Steel rolling processing	Coke oven gas, blast furnace gas, converter gas, electricity	–	About 8%
Other auxiliary processes	Coke oven gas, blast furnace gas, converter gas, electricity	–	About 58%

Note The self-generated electricity and heat within the boundaries of the enterprise is not taken into account process occupies a smaller share, around 1%. Some steel enterprises have merged the sintering and pelletizing processes, so that the energy consumption and carbon emissions of the pelletizing process are not calculated separately.

The blast furnace ironmaking process is a primary source of energy consumption and carbon emissions in steel production enterprises. During the ironmaking process, anthracite coal and coke are utilized as fuels and reducing agents, while blast furnace gas is generated, with a portion being used internally and the remainder being supplied to other processes. Carbon emissions from this process account for approximately 25%.

The converter steelmaking process serves as a net energy output process because the output of converter gas makes the emissions from this process to be negative.

The steel rolling process, which is the final stage of steel processing, primarily consumes surplus gas transferred from other processes, contributing to carbon emissions of about 8%.

Other auxiliary processes mainly include the self-supplied power plants of steel enterprises, which utilize coke oven gas, blast furnace gas, and converter gas from other processes to generate electricity and steam for the use in other production stages.

c. Application of Key Carbon Reduction Technologies

The carbon reduction technologies in the steel industry mainly include energy-saving technologies and process innovation technologies. Energy-saving technologies focus on continuously developing and applying advanced energy-saving techniques based on the present performance to reduce energy consumption during the production process, thereby reducing carbon dioxide emissions. Process innovation technologies involve a profound transformation of steel production techniques, so as to completely eliminates carbon dioxide emissions from the production process and achieve carbon neutrality.

(i) Typical Energy-Saving Technologies in the Steel Industry

The steel production process is lengthy and complex. The typical energy-saving technologies widely used in various processes are shown in Table 4.

(ii) Recovery and Utilization of Waste Heat from Blast Furnace Slag Flushing Water

The high-temperature waste heat in the steel production process is relatively easy to recover, and most of it has already been recovered through the technological transformation for energy-saving and consumption-reduction. However, the recovery and utilization of low-temperature waste heat are limited. For example, the waste heat of blast furnace slag flushing water is mostly wasted. Low-temperature waste heat accounts for approximately 35% of the total waste heat, indicating a substantial potential for recovery in the steel industry.

In the blast furnace ironmaking process, the temperature of the generated slags is approximately 1000 °C. Currently, most ironmaking enterprises handle these slags by rapidly cooling them into water slags using high-velocity water flow provided by a slag flushing pump in a slag flushing tank, which are then utilized for cement

Table 4 Typical energy-saving technologies used in various processes of steel industry

Process	Name of the energy-saving technologies
Coking	High temperature and high pressure dry quenching technology Coal moisture control technology
Sintering	Raw material layer sintering technology Low temperature sintering Technology for reducing sintering air leakage rate Sintering process for small pellets
Ironmaking	Energy saving technology for rotary cutting top burning hot blast stoves with high-temperature Double preheating technology for blast furnaces and hot blast stoves Energy saving technology for dehumidification of blast furnaces and blasts
Converter furnace	Optimized scheduling for steelmaking and continuous casting Highly-efficient continuous casting technology
Electric furnace steelmaking	Optimized power supply technology for electric arc furnaces Pre-treatment technology for scrap steel processing
Steel rolling	Online heat treatment technology for medium and thick plates Low-temperature rolling technology Regenerative combustion technology for rolling mill heating furnaces

production. Typically, around 0.3 tonnes of slags are produced for every tonne of iron, and each tonne of slags can generate 5–10 tonnes of slag flushing water at temperatures between 80 and 95 °C. To ensure the effective recycling of the slag flushing water, it must be introduced into an air cooling tower after sedimentation and filtration, where it is cooled down to below 50 °C before being recycled for slag flushing process. This method results in a significant loss of heat in the air cooling tower, leading to energy waste. By effectively utilizing this waste heat for purposes such as heat and hot water supply, we can conserve energy, reduce operational costs, and protect the environment by minimizing thermal pollution.

(iii) Waste Heat Recovery in Sintering Process

The energy consumption of the sintering process accounts for approximately 12–15% of the total energy consumption in steel production, making it the second highest, behind the ironmaking process only. Additionally, the waste heat emitted from this process accounts for 49% of the total thermal energy. Therefore, recovering and utilizing this waste heat is distinctly important. In our country, the recovery of sensible heat from sinter ores is restricted by technological bottlenecks, resulting in a low recovery rate. Therefore, the potential benefits of waste heat utilization are substantial.

The waste heat recovery and utilization technology for annular coolers can recover the hot air discharged from the first and second sections of the wind boxes of the sintering annular coolers as the heat source for the waste heat boilers. The hot air can then drive the steam turbine generator sets for electricity generation. At the same time, the low-temperature steam can also be extracted to preheat the materials to be sintered, increasing the material temperature so as to reduce the energy consumption of sintering. Through comprehensive utilization, the sintering waste heat can be utilized to the greatest extent.

(iv) Energy Conservation and Emission Reduction in Poly-generation Systems for Steel, Chemical Products, and Electricity

In the steel production process, a significant amount of secondary energy is generated, including by-products such as coke oven gas, blast furnace gas, and converter gas, which contain a large amount of CO, CH_4, and H_2. These gases serve as excellent feedstock for chemical production. Currently, the common practice involves directing the coal gas to self-owned power plants for electricity generation and heat supply. For instance, utilizing by-product gases for the poly-generation of chemical products, electricity and steam can enhance energy utilization efficiency, representing an effective approach for steelmakers to achieve energy conservation and emission reduction.

The steelworks utilize all by-product gases for electricity generation, leading to resource devaluation and waste. The coke oven gas has a high content of hydrogen and hydrocarbon which can be utilized for the production of chemical products such as hydrogen, methanol, ethylene glycol, natural gas, and synthetic ammonia. Converter gas, high in carbon monoxide content and low in impurities, can be processed through adsorption technology to isolate the carbon monoxide which is essential for producing important chemical feedstocks such as methanol and sodium formate.

Studies indicate that, compared to the steel-electricity model where all by-product gas is utilized for power generation, the three poly-generation models, namely the co-generation of steel-electricity-methanol, the co-generation of steel-electricity-hydrogen, and the co-generation of steel-electricity-methanol-hydrogen, can enhance energy utilization efficiency by over 10%. Moreover, their economic benefits are significantly higher compared to the steel-electricity model.

(v) Reducing the Iron-to-Steel Ratio and Increasing the Scrap Steel Ratio

In 2019, China's crude steel output reached 996 million tonnes, accounting for 53.3% of the global steel output (1.869 billion tonnes); pig iron output was 809 million tonnes, representing 64.2% of the world's pig iron output (1.26 billion tonnes). Based on these figures, the iron-to-steel ratio in China's steel industry was 0.812, which was 15.7% higher than the global average (including China) of 0.685. Iron ore and scrap steel are the primary ferrous feedstock, and a high iron-to-steel ratio indicates a low scrap steel ratio.

As a carrier of energy, scrap steel does not need to be reduced into molten iron in a blast furnace like iron ore. Consequently, the comprehensive energy consumption per tonne of steel produced using a short-process method with scrap steel as the feedstock accounts for approximately 20% that of the "blast furnace-converter" long-process method with iron ore as the main feedstock. In 2019, the energy consumption per tonne of steel for major domestic steelmakers was as follows: 552.06 kg of standard coal for the long process and 106.77 kg of standard coal for the short process. For the long process, since 90% of energy consumption occurs before the blast furnace operation, implementing appropriate technological measures to increase the scrap steel ratio and reduce the iron-to-steel ratio in the long process could significantly lower the total energy consumption and carbon dioxide emissions of China's steel industry.

With the development of China's economy and the rise of its steel industry, the accumulated amount of domestic scrap steel has been steadily increasing. In 2015, China's annual output of scrap steel was 150 million tonnes; in 2019, the volume of scrap steel resources reached 240 million tonnes. According to predictions from the Chinese Academy of Engineering, by 2025, China's steel stock will reach 12 billion tonnes, and the volume of scrap steel will exceed 280 million tonnes; by 2030, the volume of scrap steel resources is expected to surpass 330 million tonnes. There will be ample resources available in the future to increase the scrap steel ratio and reduce the iron-to-steel ratio, thereby significantly reducing the total energy consumption and carbon dioxide emissions of the steel industry.

In summary, considering the process characteristics and the future development trends of China's steel industry, the adoption of new processes that emphasize high scrap steel ratios and efficient smelting in converters will be a primary direction for reducing carbon emissions in the steel sector.

(vi) Hydrogen Based Steelmaking

The new low-carbon or carbon-free ironmaking processes represented by hydrogen-based metallurgy are major topics for the source control of carbon emissions in the steel industry.

Green hydrogen produced by renewable electricity can replace the coke used in conventional ironmaking process. In conventional ironmaking process, the carbon in coke reacts with the oxygen in iron ore and that generates carbon dioxide, which is used as a reducing agent. By replacing coke with hydrogen, the hydrogen will react with the oxygen in iron ore and it will generate water, thus achieving zero emissions in the ironmaking process.

Currently, "green steel" projects that utilize hydrogen as a substitute for coke are being implemented both at home and abroad. SSAB (Swedish Steel), Vattenfall (a Swedish state-owned power company) and LKAB (a Swedish state-owned iron ore producer) have taken a decisive step towards fossil-free steelmaking with the official launch of the fossil-free sponge iron production plant under the HYBRIT project. The HYBRIT project aims to replace coking coal, which is conventionally used in ore-based steelmaking, with fossil-free electricity and hydrogen. The hydrogen-rich carbon recycling blast furnace pilot project of Bayi Iron and Steel under China Baowu Steel Group has started its construction. This pilot project is intended to further transform the way of extracting smelting kinetic energy, from burning carbon in the past to burning hydrogen, so as to further reduce carbon usage and carbon emissions and even achieve zero emissions.

Although hydrogen-based metallurgy is currently still in the experimental stage and has not yet met the conditions for commercialization and industrialization, it is foreseeable that in order to achieve deep decarbonization in the iron and steel smelting field and achieve carbon neutrality in the steel industry, it is imperative to make breakthroughs in key low-carbon technologies.

2.2 *The Characteristics of Emissions and the Application of Key Carbon Reduction Technologies in the Petrochemical Industry*

a. Emission Characteristics

The petrochemical industry is characterized by extensive production lines, wide coverage, diverse product categories, and high relevance with other sectors. This industry is not only a significant consumer of energy but also a major contributor to carbon emissions. According to the national greenhouse gas inventory, in 2014, the carbon dioxide emissions of China's petrochemical industry were approximately 920 million tonnes, accounting for 8% of the country's total CO_2 emissions that year, which could significantly impact the achievement of local targets with regard to carbon emission intensity and carbon peaking.

Based on data from the *China Statistical Yearbook (2020)*, in 2018, the petroleum, coal, and other fuel manufacturing industries accounted for 9.22% of the total energy consumption in the national industrial sector, ranking the fourth in the manufacturing industry with a share of 11.09%.

The petrochemical industry refers to enterprises that use petroleum and natural gas as the main feedstocks to produce petroleum products and petrochemical products, including refineries, petrochemical plants, petroleum chemical fiber plants, etc., or enterprises formed by the collaboration of the above-mentioned plants.

The categories of emission sources in the petrochemical industry are as follows.

A. CO_2 emissions from fuel combustion, mainly referring to the CO_2 emissions generated by the direct combustion of fossil fuels in atmospheric pressure furnaces, vacuum furnaces, flares, cracking furnaces, steam boilers, and other industrial kilns.
B. CO_2 emissions from flare combustion, mainly including the combustion of flare gas under normal operating conditions and the combustion of flare gas caused by accidents.
C. CO_2 emissions in the industrial production process, mainly referring to the direct CO_2 emissions generated during the production process of catalytic cracking units, catalytic reforming units, catalyst coking and regeneration in other production units, hydrogen production units, coking units, petroleum coke calcination units, oxidized asphalt units, ethylene cracking units, ethylene glycol/ethylene oxide production units, and other product production units.

For example, catalytic cracking is one of the processes in petroleum refining, where heavy oil undergoes a cracking reaction under the action of heat and catalysts, and then is converted into cracked gas, gasoline, diesel, and so on. In the catalytic cracking process, the by-product coke produced from the reaction deposits on the surface of the catalyst, which easily causes the catalyst to lose its activity. Generally, enterprises adopt the method of continuous coking to remove the coke on the surface of the catalyst. For continuous coking, the tail gas generated by coking may be directly discharged, which includes the direct emission of CO_2.

D. Amount of CO_2 recovered and utilized: This includes the portion of CO_2 that enterprises recover from fuel combustion or industrial production processes for their own use as production materials, as well as the portion supplied to other organizations as products. The amount of CO_2 recovered and utilized can be deducted from the total emissions of the enterprises.
E. Implicit CO_2 emissions from net purchased electricity and heat.

b. Application of Key Carbon Reduction Technologies

Carbon emission reduction technologies in the petrochemical industry mainly fall into three categories: source substitution, reduction in process, and comprehensive utilization.

Source substitution: Low-carbon new energy technologies are used to replace traditional fossil fuel technologies, thereby reducing the use of materials that emit greenhouse gases. Examples include technologies for the development and utilization of unconventional natural gas, renewable energy technologies, and non-food biofuel production technologies.

Reduction in process: Energy-saving, emission reduction, and efficiency improvement technologies are adopted to reduce the emissions of greenhouse gases. The main technical measures include the utilization of waste heat and pressure, thermal combination of installations, improvement of the energy efficiency of the installations themselves, and process improvement to boost production efficiency.

Comprehensive utilization: Light hydrocarbon recovery, the capture, utilization and storage of carbon dioxide, and carbon chemical technology are employed to recover and comprehensively utilize the emitted greenhouse gases.

According to the *National Research Progress Report on Addressing Climate Change*, in China's petrochemical industry, by promoting the ethylene production process in which naphtha and light diesel are used as the feedstock and the ethylene cracking furnace air preheating technology, large-scale operation of ethylene cracking furnaces can be achieved. By 2030, the absolute CO_2 emission reduction potential will be 34 million tonnes. By recovering and utilizing the waste heat of flue gas and low-temperature heat energy, and improving the recovery rate of flare gas, the absolute CO_2 emission reduction potential will be 15 million tonnes by 2030. Through the application of the CCS (Carbon Capture and Storage) technology in the petrochemical industry, by 2030, the absolute CO_2 emission reduction potential will be 14 million tonnes.[1]

(i) Source Substitution

Ethylene is one of the most important products in the petrochemical industry. There are mainly five production processes for producing ethylene, namely steam cracking, coal-to-olefins, methanol-to-olefins, Fischer–Tropsch synthesis, and refinery tail gas. Using naphtha and light diesel as the production feedstock instead of coal-to-olefins, and using dry gas instead of fuel oil as the fuel for heating furnaces can reduce carbon dioxide emissions from the source.

(a) New Technologies for Steam Cracking Units

Steam cracking is the main method for producing low-molecular olefins such as ethylene and propylene. The ethylene steam cracking units are the most developed facilities in the chemical industry and also one of the largest carbon dioxide emission sources in the entire petrochemical industry chain. The revolutionary new technology of replacing traditional fossil fuels with renewable electricity to heat cracking furnaces, is expected to achieve an emission reduction of up to 90%.

(b) Production Technology of Fuel Ethanol

Fuel ethanol is generally obtained through processes such as biological fermentation or enzymatic hydrolysis with biomass as the feedstock. It is a clean fuel with high octane and also an excellent fuel improver. As a renewable fuel, fuel ethanol will fundamentally reduce the combustion emissions of fossil fuels.

(ii) Reduction in Process

(a) Recovery of Process Waste Heat

The petrochemical industry consumes a large amount of energy and has abundant thermal resources. Technologies such as distillation energy conservation, high-efficiency heat exchangers, thermal combination among installations, recovery and utilization of high and low temperature heat, as well as energy integration and recovery are adopted to reduce energy consumption or improve energy utilization efficiency.

The ethylene cracking furnace are major energy consumers in the ethylene plants, with the fuels consumed accounting for 60–80% of the total energy consumption of the ethylene plants. On the premise of meeting the production needs of the cracking furnaces in ethylene plants, the technology of using the waste heat within the ethylene production unit area itself to preheat air can significantly reduce the energy consumption of ethylene cracking furnaces. Applicable low-temperature waste heat sources includes the medium and low-pressure steam, steam condensate, quench water and heat tracing water in the ethylene complex unit of the olefin plants. This technology is applicable to the burners at the bottom of cracking furnaces. By installing an air preheater at the bottom burner of the cracking furnace to heat the inlet air, the temperature of the incoming air will be higher, thus reducing the fuel consumption of the cracking furnaces and the CO_2 emissions.

(b) Complete Set of Reactive Distillation Technology

The complete set of reactive distillation technology is a process that separates products by means of distillation while reactions are occurring simultaneously. This technology offers advantages such as high conversion rates, excellent selectivity, and low energy consumption, making it widely applicable in processes such as esterification, hydrolysis, transesterification, and polymerization. Compared to the processes where reaction and separation are carried out independently, the conversion rate can be enhanced by 30–50%; catalyst utilization can be improved by 80–110%; selectivity can be improved by 10–40%; energy consumption can be reduced by 20–50%; and production capacity can rise by 20–40%.[1]

(c) Recovery of Flare Gas

In petrochemical plants, flare systems are adopted to handle the combustible gases and combustible toxic gases discharged during the production process and in accidents. The gases fed into the flare system are burned off with no additional use, which not only wastes materials but also increases carbon dioxide emissions. Recovering and utilizing flare gas can reduce energy consumption and carbon emissions. Flare gas can be recovered in two ways, online recovery and gas holder recovery. Online recovery involves sucking, compressing and transporting flare gas to the fuel system by compressors. Gas holder recovery entails storing the flare gas in the gas holder first and then transporting it to the fuel gas system.[2]

(d) Coke Formation Inhibition Technology

When ethylene is produced by steam cracking, coke will be formed on the inner walls of the furnace tubes in the radiant sections of the cracking furnaces and on the inner walls of the cased tubes of the quench boilers. The coke formed under such high-temperature conditions acts as a poor thermal conductor, which will increase the heat transfer resistance of the furnace tubes, reduce thermal efficiency and increase energy consumption. Additionally, it may block the furnace tubes, deteriorate the material properties of the furnace tubes and affect normal production. To ensure production, decoking operations need to be carried out continuously; however, decoking will generate carbon dioxide emissions.

The methods for coke formation inhibition can be categorized as follows: ① Improving tube materials: Utilizing new anti-coking tube materials such as alloy tubes and ceramic tubes of high performance. ② Enhancing cracking feedstock quality: Pre-treating the feedstocks and improving their quality can reduce the aromatic content and increase the hydrogen content, thus effectively inhibit coke formation. The cracking feedstock pretreatment methods adopted in the industry mainly include processes such as hydrotreating and aromatic extraction. ③ Adding coke formation inhibitors: Adding an appropriate amount of coke formation inhibitors to the cracking feedstocks or dilution steam, thus reducing the coke formation rate and changing the coke structure through chemical reactions; or catalyzing the gasification of the coke to reduce coke formation and prolong the operation cycle of the cracking furnaces. ④ Increasing heat transfer: Using enhanced heat transfer tubes to improve thermal efficiency and shorten the residence time of materials in the tubes, thus slowing down coke formation in the furnace tubes.[1]

(e) Key Technologies for Full-Process Intelligent Control of Large-Scale Petrochemical Facilities and Smart Energy Systems

The smart energy system deeply integrates the Internet, big data, and artificial intelligence with the petrochemical production processes. It addresses technical challenges such as multi-variable, nonlinear, strong coupling, pure delay, coexistence of intermittent and continuous control, multi-constraint and multi-target regulation in petrochemical production facilities. By enhancing the level of intelligence of the petrochemical production facilities, the system can achieve a stability rate of 100%, and the energy consumption can be reduced by 0.5–10%.[1] The smart energy system provides petrochemical enterprises with integrated and visualized energy management services with regard to big data processing capability in areas such as water, electricity, gas, and heat. It helps users with the collection, storage, display, computation, and analysis of energy data, while automating routine energy management tasks such as meter reading and anomaly reporting according to established schedules, thereby improving work efficiency. Additionally, through the accumulation of big data, it can uncover energy-saving potential and further reduce energy consumption.

(iii) Comprehensive Utilization

There are numerous approaches for the development and application of carbon dioxide in the petrochemical industry. It serves as a synthesis feedstock for various chemical products. By leveraging carbon dioxide resources, reactions can occur in the presence of catalysts with hydrogen, water, oxygen and ammonia to produce methanol, dimethyl ether, low-carbon olefins, low-carbon alcohols, mixed fuels, synthetic esters, carboxylic acids, synthetic amines, formic acid and its derivatives, aldehydes, among others. Additionally, carbon dioxide can be converted into carbon monoxide syngas.

The single-pass yield of converting carbon dioxide into gasoline is 26%. Synthetic fuel facilities have been established abroad to produce clean diesel, naphtha, and paraffin from methane and carbon dioxide. Japan has developed a new two-step efficient hydrocarbon synthesis process using carbon dioxide and natural gas as raw materials, with a single-pass yield of 80%. The synthesized hydrocarbons can be simply separated to obtain gasoline, diesel, kerosene and so on. In recent years, research on the CO_2 hydrogenation to produce hydrocarbons has also been carried out in China. The carbon monoxide and hydrogen produced by the reforming of carbon dioxide and methane are precisely in the ratio suitable for the synthesis of oil. Additionally, carbon dioxide can be used to prepare high molecular compounds such as carbonates, polyureas, polyurethanes, polyketones, polyethers, polyketone ether esters, and liquid crystal polymers.[2]

(a) Carbon Capture, Utilization, and Storage (CCUS) Technology

The CCUS technologies developed by China Petroleum and Chemical Corporation, including innovative low partial pressure carbon dioxide recovery technology, are extensively applied in the capture of carbon dioxide from gases such as natural gas and syngas. The captured carbon dioxide is utilized in the production of urea, methanol, syngas, food-grade carbon dioxide, and EOR, etc.[3]

(b) Steam Methane Reforming (SMR) Hydrogen Production Technology Equipped with Carbon Capture and Storage (CCS)

Steam methane reforming (SMR) is the most established hydrogen production technology in the industry, accounting for approximately 70% of global hydrogen output. However, it has the drawback of emitting greenhouse gas carbon dioxide. By integrating SMR with CCS, carbon dioxide generated from the reforming process can be captured and sequestered using vacuum pressure swing adsorption method.

2.3 The Characteristics of Emissions and the Application of Key Carbon Reduction Technologies in the Chemical Industry

a. Emission Characteristics of the Chemical Industry

The chemical industry is a major consumer of the three major energy sources: oil, coal and natural gas. Energy serves as both the power and fuel for the chemical industry and is also a chemical feedstock. The contribution of fossil feedstocks extends beyond the energy released through combustion; it is a coupled process of chemical reactions and energy conversion, with a large amount of carbon entering into the products through chemical reactions. The difference between the carbon entering the boundary of the chemical industry and the carbon taken out by the products represents the lost carbon, which, when converted into carbon dioxide, reflects the carbon dioxide emissions of the chemical industry.

The chemical field encompasses a wide variety and large number of sectors, and there are significant differences among them. Due to the complexity of production technologies and the diversity of end products, assessing carbon emissions in the chemical industry poses considerable challenges. Even for the same product, variations in processes can lead to substantial differences in the carbon emission intensity per unit of product.

The primary categories of emission sources within the chemical industry are as follows.

(i) Emissions from Fuel Combustion

Emissions from fuel combustion mainly refer to the CO_2 emissions generated in the complete combustion of fossil fuels with oxygen in various types of stationary or mobile combustion equipment (such as boilers, burners, turbines, heaters, incinerators, kilns, furnaces, ovens, internal combustion engines, etc.). This part of fossil fuels refers to the portion used as fuels in combustion equipment, excluding the part of fossil fuels produced as by-products or combustible waste gases in industrial production processes that are recovered and burned as energy.

(ii) Emissions from Industrial Production Processes

(a) CO_2 Emissions from Raw Material Consumption

CO_2 emissions from feedstock consumption mainly refer to the CO_2 emissions generated by the use of fossil fuels and other hydrocarbons as feedstocks, including the CO_2 emissions generated after the vented gases are treated with flares. The chemical industry is characterized by the fact that fossil fuels and other hydrocarbons are often used as feedstocks. For instance, CH_4 (methane), C_2H_4 (ethylene), C_2H_2 (acetylene), and CH_2=CH–CH_3 (propylene) are the main direct feedstocks for organic chemicals. The use of these hydrocarbons as feedstocks leads to CO_2 emissions.

(b) CO_2 Emissions from the Use of Carbonates

CO_2 emissions from the use of carbonates mainly refer to the CO_2 emissions generated during the use of carbonates (for example, limestone and dolomite are used as feedstocks, fluxes or desulfurization agents).

(c) N_2O Emissions from Processes such as Nitric Acid and Adipic Acid Production

In the production of nitric acid, the catalytic oxidation of ammonia at high temperature results in by-product N_2O. In the production of adipic acid from cyclohexanone/cyclohexanol mixture which is oxidized by nitric acid, a by-product of N_2O is generated. According to the Second Assessment Report of the IPCC, the warming potential of 1 tonne of N_2O over a 100-year time scale is equivalent to that of 310 tonnes of CO_2.

(iii) The Amount of CO_2 Recovered and Utilized

The amount of CO_2 recovered and utilized primarily refers to the part of carbon dioxide recovered by the reporting entity from fuel combustion or industrial production processes and subsequently supplied as a product to other organizations, which should be deducted from the total emissions. This does not include the part of CO_2 that is recovered and utilized on-site by the enterprise itself.

In cases where some chemical companies, such as nitrogen fertilizer plants, recover CO_2 and sell it as a product, then this part of CO_2 is not included in the company's total emissions.

(iv) Emissions from Net Purchased Electricity and Heat

The emissions of this part will be calculated based on the net purchased amount of the enterprise.

(v) Other Greenhouse Gas Emissions

Chemical enterprises have a wide variety of products, and the production processes are complex and diversified. This requires enterprises to identify the emission sources of each process according to the actual production conditions.

b. Application of Key Carbon Reducution Technologies

Chemical enterprises are actively conducting researches on energy substitution, aiming to replace high-carbon energy with low-carbon renewable energy. Given the unique characteristics of the chemical industry, waste materials generated from sub-sectors such as coal chemical sector and calcium carbide are often utilized as feedstocks in the production process of other sub-sectors, like methanol. Therefore, it is essential to develop an industrial chain within the chemical sector, applying the principles of a circular economy to the research and development of exhaust gas utilization devices. This involves the invention of high-tech devices for exhaust gas recovery and utilization, and efficient production of by-products, thereby establishing an industrial chain and a circular economy that can facilitate the recycling of energy and resources while reducing carbon emissions.

Additionally, several new chemical processes that utilize carbon dioxide as a feedstock are rapidly advancing, such as the production of urea, salicylic acid, methanol, and clean fuels using the carbon dioxide generated by ammonia synthesis plants. By adopting new technologies that emerged in the chemical industry, various stages or specific links of energy processing can be optimized fundamentally, such as enhancing the separation and cracking capabilities, and reducing reaction temperatures, thus improving production efficiency, reducing energy consumption and carbon emissions.

According to the "National Research Progress Report on Addressing Climate Change", by promoting and implementing the energy-saving transformation of electrolyzers in the caustic soda industry and the advanced ion exchange membrane technology for caustic soda, the absolute potential for CO_2 emission reduction will reach 35 million tonnes by 2030; by adopting advanced ammonia-soda production processes and combined soda production processes in the soda ash industry, the absolute potential for CO_2 emission reduction will reach 3 million tonnes by 2030; by carrying out the transformation of advanced fully enclosed calcium carbide furnaces and open calcium carbide furnaces, the absolute potential for CO_2 emission reduction will reach 7 million tonnes by 2030; through the replacement of coal-based ammonia synthesis plants with large-scale natural gas, technologies for the comprehensive transformation of ammonia synthesis process, and energy-saving technologies for molecular sieves in the ammonia synthesis loop, the absolute potential for CO_2 emission reduction would reach 54 million tonnes by 2020 and 77 million tonnes by 2030.[1]

(i) New Processes for Carbon Reduction and Energy Saving

(a) Natural Gas-Based Ammonia Production

The energy consumption of the synthetic ammonia industry accounts for approximately 25% of the total energy consumption in China's chemical industry, ranking first among the top five energy-intensive industries in the chemical industry. Ammonia production is at the upstream of the nitrogen fertilizer industry. Since it involves the conversion process of fuel and energy, it accounts for the main sources of energy consumption and CO_2 emissions in nitrogen fertilizer production. The production processes for synthetic ammonia can be categorized into gas-based, coal-based, and oil-based process routes, depending on the primary feedstocks used. Given the feedstock structure, the emissions from natural gas-based ammonia production are relatively low, accounting for 12% of the total CO_2 emissions from the synthetic ammonia industry. Therefore, it is advisable to vigorously promote natural gas-based synthetic ammonia production while carrying out co-generation of heat and power, multi-stage utilization of steam, and waste heat recovery to reduce carbon emissions.

(b) Chlor-Alkali Industry

The chlor-alkali industry employs an electrolysis process that is highly energy-intensive, with carbon emissions primarily coming from electricity consumption. Currently, a number of energy-saving and emission-reduction new technologies are being promoted within the sector, including membrane-based brine purification, membrane denitrification, bipolar ion-exchange membrane electrolyzer with membrane (zero) electrode distance, dry-process acetylene, low-mercury catalysts, large-scale polymerizers with a length of over 100 m, utilization of waste heat from hydrogen chloride synthesis, deep desorption of hydrochloric acid, treatment of PVC polymerization mother liquor, and comprehensive utilization of calcium carbide slag. Meanwhile, the research and development of some international cutting-edge technologies, including domestic ion-exchange membrane manufacturing, oxygen-depolarized cathode (ODC) electrolyzers, and plasma pyrolysis of coal powder for acetylene production are also progressing steadily.[1]

The ion-exchange membrane method for alkali production offers advantages such as low energy consumption, low costs, low pollution from the three wastes, and ease of operation. The ion-exchange membrane alkaline solution contains only trace amounts of salt. No desalting is required during its entire evaporation and concentration process, even when producing solid caustic soda with a content of 99%, greatly simplifying the process and the equipment. The concentration of the solution is high, generally ranging from 30 to 33%, which is highly beneficial for the concentration and evaporation of the solution, leading to substantial reductions in the amount of steam required for concentration and, consequently, reducing energy consumption.

The technology of recovering hydrogen from the tail gas of vinyl chloride distillation refers to the recovery of hydrogen from the tail gas of vinyl chloride distillation by using the pressure swing adsorption technique after recovering acetylene and vinyl chloride from the tail gas. This technology is applicable to the transformation projects of enterprises that use the calcium carbide process to produce PVC. In the transformation project, vinyl chloride and acetylene are recovered from the distillation tail gas by means of pressure swing adsorption. This technology can replace the original natural gas-based hydrogen production installations,[1] thus avoiding CO_2 emissions caused by hydrogen production from natural gas.

(c) Soda Ash Industry

The ammonia-soda production process can be replaced by the combined soda production process. By leveraging the characteristics that ammonia synthesis plants can produce a by-product of CO_2, the two major systems of ammonia production and soda production can be integrated into a single continuous production line. The combined soda ash production process is characterized by high feedstock utilization, low energy consumption, and minimal discharge of waste residues and waste liquid. Furthermore, the innovative shift gas-based soda ash production technique in the combined soda production process represents a new technology with Chinese characteristics, which transforms the traditional three-tower soda production process into a single-tower production process, and shifts from internal to external heat exchange, thereby reducing resistance and conserving energy.

(d) Urea Synthesis Technology with High Efficiency and Low Energy Consumption

This technology employs a urea synthesis high-pressure loop in a fully condensed reactor and a two-stage process flow. A simple and straightforward medium-pressure system is set up, which reduces the load of the high-pressure stripping tower and the consumption of medium-pressure steam. This process consumes less energy than the traditional aqueous circulation urea production plants and the CO_2 stripping urea production plants.

(ii) Control of Nitrous Oxide Emissions from Industrial Production Processes

Controlling the output of nitric acid and adipic acid can mitigate excess production capacity, thereby reducing nitrous oxide emissions generated from the production of nitric acid and adipic acid at the source. Enhancing the production processes of existing nitric acid facilities and promoting the use of secondary treatment methods to control nitrous oxide emissions; the newly constructed nitric acid production facilities should adopt tertiary treatment methods for nitrous oxide decomposition. For existing adipic acid production facilities, promoting the adoption of catalytic decomposition techniques while encouraging the application of thermal decomposition technologies in newly constructed adipic acid installations.[1]

(iii) Carbon Dioxide Capture Technologies

Large-scale coal to liquid chemical projects mainly employ the low-temperature methanol washing process to remove CO_2. If a CO_2 product tower is set up, CO_2 with a volume fraction of over 98% can be obtained.

Low-temperature methanol washing technology uses cold methanol as the absorption solvent. This technology takes advantage of the excellent property of methanol that it has high solubility for acidic gases (such as CO_2, H_2S, and COS) at low temperatures; therefore, the removal of acidic gases from the feed gas can be achieved through this physical absorption process. Low-temperature methanol washing technology is recognized in China and abroad as the most economical and effective gas purification technology, with features that cannot be replaced by other desulfurization and decarbonization technologies.

The Catacarb process for decarbonization has been widely adopted by ammonia synthesis enterprises. In addition, there are also other physical absorption methods, such as the multi-amine method for decarbonization, the new non-carcinogenic methylenediamine (NCMA) process for decarbonization, and polyethylene glycol dimethyl ether (NHD), which are applied in various industrial decarbonization installations.

The new NCMA decarbonization technology uses several active amines with steric hindrance effects to form a compound amine solution (MA solution) which is used as the decarbonization solution. This new technology is developed on the basis of the methyldiethanolamine (MDEA) decarbonization technology, aiming at making up the deficiencies of the MDEA decarbonization technology. It has the dual properties of physical absorption and chemical absorption, and boasts good energy-saving effects and a relatively high purification degree.[1]

NHD is a non-toxic, non-hazardous and solvent corrosion resistant polymer compound with good thermal resistance and good chemical stability. It demonstrates a strong absorption capacity for compounds such as hydrogen sulfide, carbon dioxide, carbonyl sulfide, and thiols in combustible gases like refinery gas, coal chemical synthesis gas, natural gas, and oil field gas. Once the NHD solvent reaches saturation, the adsorbed substances can be removed through vacuum depressurization and temperature rise, thereby facilitating the recycling and regeneration of the solvent.[1]

2.4 The Characteristics of Emissions and the Application of Key Carbon Emission Reduction Technologies in Cement Industry

Cement is an important basic raw material for China's economic and social development, and is widely used in civil engineering, water conservancy, national defense and other fields, playing an important role in improving people's livelihood, promoting national economic growth and safeguarding national security. During the 13th Five-Year Plan period, the building materials industry in our country underwent a transformation and upgrade, achieving substantial progress in technological innovation and advancement, development enhancement, acceleration of energy conservation, emission reduction, and green development. However, the cement industry is still an energy-intensive and resource-intensive industry that contributes a large amount of carbon emissions. It faces formidable challenges in emission reduction, making it a key focus for the development of low-carbon and energy-efficient technologies.

a. Emission Characteristics

Cement production is divided into three main processes:

(i) Raw materials preparation: This involves crushing the limestone raw materials, argillaceous raw materials and a small amount of corrective raw materials first, then mixing them in a certain proportion and grinding them into fine powder. The powder is then blended into raw materials with appropriate composition and uniform quality.
(ii) Clinker calcination: The raw materials are calcined in cement kilns until partially melted to obtain Portland cement clinker with calcium silicate as

the main constituent. The carbon dioxide generated in this process accounts for 62% of the total emissions in the cement industry. In addition, the emissions from the combustion of fossil fuels in the cement industry and those from electricity consumption in cement production account for approximately 35% and 3% of the total emissions respectively.[1]

(iii) Cement grinding and delivery: Appropriate amounts of gypsum, admixtures or additives are added to the clinker and grinded together to make cement, which is then packaged and delivered.

The highest carbon emissions are from carbonate decomposition. In this process, carbonates such as calcium carbonates contained in the primary raw material limestone, are decomposed to produce carbon dioxide during clinker calcination. The amount of carbon dioxide generated in this process is determined by the carbonates contained in raw materials used in the production of clinker, such as limestone, sand shale, and iron. Utilizing non-carbonate alternative materials can effectively reduce carbon emissions. Additionally, the cement industry's carbon emissions primarily come from the combustion of fossil fuels and the electricity used for processes such as preparation of raw materials, clinker calcination, and cement grinding. The main strategies for emission reduction involve minimizing energy input and optimizing production processes.

b. The Application of Carbon Emission Technologies

In general, the application of carbon reduction technologies in the cement industry can be categorized into three main approaches: source control, process improvement, and end-of-pipe treatment. Source control involves minimizing the consumption of raw materials (such as limestone), fossil fuels (like coal), and electricity during the cement production process. Process improvement refers to the reduction of carbon emissions through the use of advanced technologies and equipment, the enhancement of production management, the optimization of resource utilization, and the recycling of wastes. End-of-pipe treatment focuses on designing and implementing effective mitigation technologies at the final stage of the cement production process to reduce carbon dioxide emissions.

(i) Fuel and Raw Material Substitution Technologies

The Cement Sustainability Initiative (CSI) anticipates that by 2050, the cement industry will achieve a reduction of 0.79 gigatonnes of carbon dioxide emissions, with energy efficiency improvements contributing 10%, alternative fuels 24%, and clinker substitution 10%. Given that the carbon emission intensity of alternative fuels is 20–25% lower than that of coal, the use of alternative fuels can significantly reduce carbon emissions generated from fuel combustion. The research on alternative fuel technologies in China's cement industry has commenced relatively late, with the current heat substitution rate lower than 2%, indicating a heavy reliance on coal-based fossil energy for cement production. China is rich in resources for alternative fuels. The resources available for producing alternative fuels primarily include industrial and municipal solid wastes, waste plastics, municipal sludge, and waste tires, presenting substantial potential for increasing the proportion of fossil energy substitutes. Furthermore, the number of cement clinker production lines that utilize alternative fuels in China is less than 1%,[1] indicating considerable room for improvement in fuel substitution within the cement industry. According to projections by the China Cement Association, driven by the goals of carbon peaking and carbon neutrality, the heat substitution rate of alternative fuels in the cement sector can reach 6% by 2030 and 45% by 2060. Promoting the industrialization of alternative fuels is crucial for enhancing the current thermal substitution rate.

In all production processes, approximately 90% of the carbon dioxide emissions come from the decomposition of carbonate raw materials such as limestone, and carbonates primary come from calcareous materials. Typically, the production of one tonne of cement clinker requires about 1.3 tonnes of calcareous materials, which, when subjected to high-temperature decomposition, generate around 42% of the carbon dioxide. The *China's Climate Change Plan (2014–2020)* emphasizes that "the cement industry should promote the use of industrial waste residues such as carbide slag, papermaking sludge, desulfurization gypsum, fly ash, and metallurgical slag tailings, as well as non-carbonate materials like volcanic ash, as substitutes for traditional limestone raw materials". Therefore, there is great potential for raw material substitution within the cement industry, and raw material substitution should become a key focus for carbon reduction efforts. Continuous efforts should be made to strengthen research on raw material substitution and enhance substitution technologies. Currently, relatively well-established alternative materials include carbide slag, silicate slag, blast furnace slag, fly ash, coal gangue, and steel slag. These industrial waste residues contain calcium oxide but have lower carbon content, effectively reducing carbon dioxide emissions during the production process when used as substitutes for limestone and other carbonate materials.

(ii) Cement Kiln Co-processing Technology

Cement kiln co-processing technology is a well-established energy-saving and emission-reduction approach in the cement industry both domestically and internationally. This technology involves putting solid wastes that meet the requirements for entering cement kilns or that meet the entering requirements after pretreatment. During the calcination of cement clinker, these solid wastes are subject to high-temperature incineration for complete decomposition. The heat generated in this process is recovered and directly utilized for the calcination of cement, thereby reducing the consumption of coal. Additionally, the ash produced from the incineration can be used as raw materials in cement production, substituting a portion of the mineral raw materials and consequently reducing the carbon dioxide emission from the decomposition of carbonates. Essentially, cement kiln co-processing technology addresses the two aspects of fuel and raw material usage, aiming to lower the reliance on fossil fuels and carbonate materials to mitigate carbon dioxide emissions. This approach not only cuts down CO_2 emissions during cement production but also provides a safe and non-harmful solution for the disposal of solid wastes such as household garbage and sewage sludge, thus facilitating the recycling of solid wastes and alleviating the environmental burden on urban areas.

(iii) Energy-Saving Grinding Equipment and Production Technologies

In the cement production process, 60–70% of the power consumed per unit of product is used for the grinding of raw materials, solid fuels, and cement clinker. Therefore, it is essential to transform the grinding process system, which primarily relies on ball mills, by adopting equipment that is based on the "material layer crushing principle" and has superior performance, advanced structures, high energy utilization efficiency. This equipment include vertical mills, roller presses, and roller mills. By adopting energy-efficient grinding equipment and production technologies, it is possible to achieve a 30–40% reduction in energy consumption of the grinding systems, thus cutting down the power consumption per unit of cement by 20–30%, and reducing carbon dioxide emissions in the cement production process.

(iv) Waste Heat Power Generation Technology

Pure low-temperature waste heat power generation involves utilizing the exhaust gases discharged from the preheater at the kiln's tail and the grate cooler at the kiln's head during the clinker calcination process to generate electricity for cement production. This part of low-temperature exhaust gas, with its temperature being below 300 °C, accounts for approximately 33% of the total heat consumption for cement clinker sintering. Waste heat power generation saves on the purchased electricity from the grid, thus reducing carbon dioxide emissions associated with power generation.

(v) Carbon Capture Technology

Carbon capture technology involves capturing the carbon dioxide discharged from cement kilns, then compressing it into a liquid and transporting it through pipelines to deep underground for permanent storage. This technology can be categorized into oxy-fuel combustion capture and post-combustion capture.

The principle of oxy-fuel combustion technology is to use an air separation device to separate air into O_2 and N_2, and then use the separated high-concentration O_2 for clinker calcination. Replacing air with this O_2 will obtain relatively pure CO_2, which can be stored after being purified and compressed. To maintain an appropriate flame temperature, a certain amount of flue gas rich in CO_2 is recycled. Due to the oxy-fuel combustion in cement kilns, the production capacity of the kilns can be improved and the proportion of alternative fuels used can be increased.

Post-combustion capture refers to the process of capturing or separating CO_2 from flue gases after combustion. This technology represents the most effective decarbonization method for coal-fired and gas-fired units with large installed capacity, applicable not only to new production lines but also to the retrofitting of existing ones. The primary flue gas decarbonization technologies include chemical absorption method, adsorption method, membrane separation method, and calcium looping method. Among them, the chemical absorption method and the adsorption method involve the use of a certain medium to absorb CO_2, thereby achieving the purpose of separating CO_2. The membrane separation method separates CO_2 by utilizing the principle that the membrane has a selective permeation for gases. Calcium looping involves the interaction of calcium oxide with CO_2-containing flue gas to produce calcium carbonate, which is subsequently decomposed during calcination to generate CO_2 and calcium oxide.

3 Transportation Area

The transportation sector, one of the major contributors to economic output, also contributes substantial energy consumption and CO_2 emissions. According to the data from the International Energy Agency, nearly 23% of global carbon emissions come from the combustion of fossil fuels in the transportation industry. Furthermore, energy consumption in this sector is growing by 43% on a yearly basis, and is expected to reach 3.26 billion tonnes of standard coal by 2035.[1] With the development of the transportation sector, carbon emissions from transportation continue to rise rapidly. Data from the "China Statistical Yearbook" illustrates that, as shown in Fig. 2, the total energy consumption in China's transportation sector has been consistently increasing, with its share in the nation's overall energy consumption being greatly increased since the beginning of the twenty-first century.

3.1 Emission Characteristics

According to the "China Statistical Yearbook (2020)", as for the data about the energy consumption in the transportation sector, the contribution of various fuels to carbon emissions in the transportation sector for 2018 was as follows: diesel > gasoline > kerosene > natural gas > fuel oil > coal. Compared to this relation of the fuels in 1997, which was coal > diesel > gasoline > fuel oil > kerosene, it changed significantly.

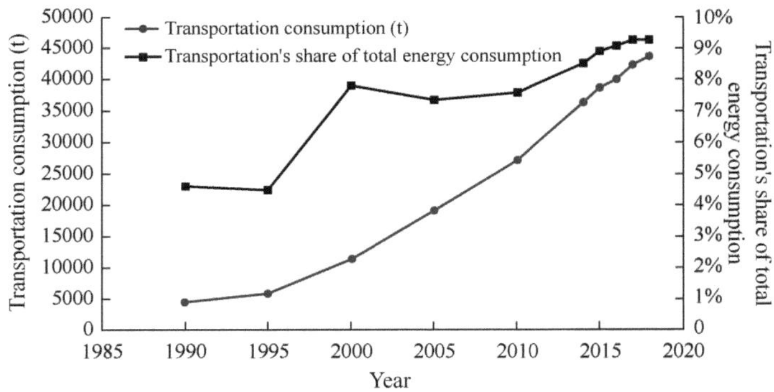

Fig. 2 Energy consumption of China's transportation sector in recent years

Over the past decade, the demand for fossil fuels in the transportation sector has increased by 25%, indicating a heavier reliance on oil and gas.

The consumption of energy in the transportation sector is closely linked to the distance and mode of transportation. In the first 15 years of the twenty-first century, the total energy consumption of global transportation sector rose by 44%, while carbon emissions increased by only 33%, highlighting improvements in energy efficiency. Notably, 72% of carbon emissions in the global transportation sector come from road vehicles, which accounts for 80% of carbon emissions in the transportation sector from 1970 to 2016.[1] With the exception of railway transportation, other modes of transportation like shipping and maritime transport have also seen increases in carbon emissions. Conversely, rail transport has experienced a decline in actual carbon emissions year-on-year due to the widespread use of electric locomotives.

In addition, economic, technological and demographic factors all have a significant impact on carbon emissions in the transportation sector. Studies have shown that the development levels of the economy, science and technology, urbanization and education, as well as per capita GDP, have a positive impact on transportation carbon emissions. The higher the levels of economic and technological development are, the larger the share of energy consumption in the transportation sector will be, and the more the carbon emissions will be. When per capita GDP reaches a certain threshold, the energy consumption in the transportation sector basically stabilizes at a certain level. Geographically, the highest transportation carbon emissions are found in developed economies and some emerging markets, while South Asia and Sub-Saharan Africa have relatively lower emissions. In 2016, the ten countries with the highest transportation carbon emissions were the United States, China, Russia, India, Brazil, Japan, Canada, Germany, Mexico, and Iran, collectively accounting for 54% of the global transportation carbon emissions in that year.[1]

3.2 *Applications of Carbon Emission Reduction Technologies*

a. Developing Clean Energy

Currently, means of transportation mainly consume gasoline and diesel. It is possible to develop and utilize ethanol, biogas, natural gas fuel, and biomass fuel as alternatives to traditional fuels. We should actively research, develop and promote technologies and products such as automatic vehicles, hyperloop, hydrogen-powered aircraft, personal electric vehicles, and electric watercraft, and gradually replace existing transportation products and technologies.

Among them, the hyperloop is a transportation system with "vacuum steel tube transportation" as its core. It adopts the "magnetic levitation + low vacuum" mode. It utilizes the low vacuum environment and supersonic shape to reduce air resistance, while reducing the frictional resistance through magnetic levitation, so as to achieve

---

[1] Wang Junwen, Ren Pingyang. The EU actively promotes emission reduction in the transportation sector [J]. Ecological Economy, 2021, 37(3): 1–4.

supersonic operation. It is characterized by ultra-high speed, high safety, low energy consumption, minimal noise and low pollution.

Hydrogen-powered aircraft uses hydrogen as the aviation fuel. Hydrogen is used as a major propulsion energy source, either for direct combustion of fuel cells in thermal (gas turbine) engines or as a component of synthetic liquid fuels for electricity generation to power electric motors. Its advantage lies in the fact that it can completely eliminate carbon dioxide emissions generated during a flight.

b. Enhancing the Efficiency of Power Combustion

The efficiency of vehicle transportation and power combustion can be enhanced through energy-saving technologies, thereby achieving the goal of carbon emission reduction. In terms of road transportation, promoting new energy vehicles and hybrid vehicles, along with applying energy-saving designs to engines, is essential. In addition, the technologies for green driving and efficient driving of automobiles should be advocated.

In the area of rail transport, it is necessary to advance the electrification of railway systems and enhance the energy storage feedback and utilization efficiency of power supply systems. Besides, we should promote lightweight and reduced weight designs for vehicles and implement efficient energy-saving designs for traction, auxiliary, and control systems.

In the shipping sector, the operational efficiency of ships can be improved through technologies such as optimized design and power upgrading. Energy-saving technologies can be adopted for cranes, like conversion from oil to electricity, potential energy recovery and supercapacitors. For port equipment, technologies such as variable frequency or direct current drive can be employed. Windproof facilities can be constructed at dry bulk terminals, and liquefied natural gas and efficient shore power technologies should be promoted at ship terminals. In terms of air transportation, it is important to use new and clean energy and lightweight substitute materials, promote the research and development of zero-emission aircraft such as hydrogen-powered ones, and vigorously developed green airports where auxiliary power units of aircraft are replaced by ground-based power units.

c. Building an Intelligent and Green Transportation Network

Vigorously developing green transportation such as new energy transportation, rail transit, public transportation, pedestrian and cycling transportation. Promoting the construction of zero-emission airports, high-speed railway networks and ports. Developing cycling infrastructure. Utilizing big data and artificial intelligence technologies to carry out dynamic and real-time carbon emission monitoring and regulation on the transportation systems. Rational design of road network structure and transportation planning based on urban traffic volume and travel distribution should be used, so as to tap the potential of the road network and optimize the structural emission reduction of transportation. For example, in terms of passenger transport, developing intelligent public transportation and advocating green travel; with regard to freight transport, promoting efficient transportation modes such as multimodal transport and drop-and-pull transport to improve freight transport efficiency.

4 Construction Area

4.1 Emission Characteristics

In 2009, data published by the United Nations Environment Programme indicated that the energy consumption of the construction industry, along with its associated production sectors, accounted for 40% of the total energy consumption of the society, and one-third of global greenhouse gas emissions were related to buildings.[1] In China, the construction sector, along with industry and transportation, represents the three sectors of the highest energy consumption and also the three largest sources of carbon dioxide emissions. According to statistics, the total energy consumption in the construction sector accounts for over 30% of China's total energy consumption, and it is projected that by 2030, this share will reach about 40%, surpassing both the industrial and transportation sectors to become the largest energy-consuming sector.[2] In summary, the carbon dioxide emissions from the construction sector in China have the following characteristics.

a. Public buildings account for a large portion of energy consumption. In 2014, the total area of public buildings was approximately 10.7 billion square meters, representing only one-fifth of the total building area. However, their energy consumption reached 235 million tonnes of coal equivalent (tce), accounting for 25% of the total energy consumption of the buildings. Meanwhile, the energy consumption intensity of public buildings has been increasing fast. From 2001 to 2014, the energy consumption per square meter of public buildings rose from 16.5 to 22.0 kgce/m^2, marking a 33% increase in energy consumption intensity. As the scale of public buildings continues to expand and their energy consumption intensity keeps rising, there remains substantial potential for emissions reduction within this sector.[2]
b. With the rapid urbanization in China, on the one hand, large-scale construction activities consume a huge amount of building materials. The production and transportation of these materials require a considerable amount of energy, resulting in a significant amount of carbon dioxide emissions, which account for a certain proportion of the total energy consumption and carbon dioxide emissions in the whole society. On the other hand, the continuous increase in building area leads to more energy use for building operations. Moreover, with the development of the economy and society, people's living standards are constantly improving. There is a rising demand for energy used by terminals such as heating, air conditioning, domestic hot water, and household appliances, which in turn contributes to more carbon emissions.

4.2 *The Application of Carbon Emission Reduction Technologies*

a. Solar Photovoltaic Power Generation Systems

Solar photovoltaic power generation systems are power generation systems that directly convert solar radiation energy into electrical energy by utilizing the photovoltaic effect of semiconductor materials in solar cells. Solar radiation energy is the huge nuclear energy released by the fusion of hydrogen and helium atoms inside the sun, representing an inexhaustible renewable energy source. As a result, photovoltaic power generation technology is increasingly applied in the construction sector, mainly through the installation of photovoltaic panel arrays on the roofs or terraces of buildings to provide energy for house heating, lighting and other electrical needs, achieving the goal of reducing the use of electricity or gas and lowering carbon dioxide emissions.

b. Green and Energy-Saving Building Materials

Green and energy-saving building materials refer to those produced by advanced manufacturing processes and those that possess characteristics such as sustainability, energy efficiency, environmental friendliness, and superior performance. The manufacturing and usage of building materials will emit large amounts of carbon dioxide. Therefore, the use of green and energy-saving building materials can help promote energy conservation and emission reduction in the construction industry. They can not only enhance the thermal insulation effect of buildings and reduce the energy loss from air conditioning, but also improve people's living and working environments. Currently, there are various types of green and energy-saving building materials, among which recyclable building materials are relatively common. They can replace materials extracted from nature, form new building components, and maintain comparable quality to traditional materials. For instance, recycled concrete can be crushed to produce recycled aggregates, recycled wood can be pulverized for the production of density fiberboard used in wood products, recycled polystyrene foam can be used as new plastic raw materials, surface coatings or paints, recycled window glass can be added to asphalt as reflective road coatings, or combined with concrete to form floor slabs and granite countertops.

c. Energy-Saving and Thermal Insulation Technology for Walls

There is a large amount of heat loss in the exterior walls of building structures. The use of green energy-saving building materials can effectively reduce heat loss. If combined with energy-saving and thermal insulation technologies for walls at the same time, the operational energy consumption of the buildings can be further reduced. Energy-saving and thermal insulation technologies for walls can be categorized into two categories: internal heat preservation and external heat preservation.

The following three methods are used for internal heat preservation and energy-saving walls: ① Pasting prefabricated heat preservation boards; ② Pasting polystyrene boards on the walls and then plastering the surface layer with gypsum; ③ After treating the interface of the base wall, directly smearing or spraying the polystyrene granule heat preservation mortar on its surface to make the surface layer of anti-crack mortar.

Compared to interior insulation walls, exterior thermal insulation and energy-saving walls makes up for many of the shortcomings, as they do not occupy interior space, have high thermal efficiency, and are also conducive to structural protection. Currently, the exterior thermal insulation method has become the main technology for energy-saving walls, mainly including the following types: ① The exterior heat preservation system for exterior walls with thin plastering of expanded polystyrene (EPS) boards. Its main components are the thin plastering layer, the EPS board heat preservation layer and the finishing coating layer. ② The exterior heat preservation system for cast-in-place concrete exterior walls with EPS boards, mainly used in the construction of cast-in-place concrete exterior walls. Its advantage lies in that it can make a tighter bond between the heat preservation board and the wall concrete, thus reducing the need for adhesive and shortening the construction period; ③ The Structural Insulated Panel System (SIPS), which utilizes polyurethane foam or polystyrene foam in the core, allowing for the selection of appropriate flat surface layers according to the conditions of both sides; ④ The exterior insulation system of inorganic insulation mortar, which is easy to operate, environmentally friendly, highly energy-saving, and easy to store. It provides good thermal insulation properties and demonstrates good resistance to impact, compression and water, thus effectively preventing problems such as wall leakage and cracking.

In addition to the above-mentioned application technologies for external thermal insulation and energy-saving walls, there is also a green ecological garden-style rooftop greening system. This system consists of specially selected plants, specially formulated nutrient soil, modified PVC root-resistant coil materials, insulation boards, waterproof layers and accessory parts. It can be applied to concrete roofs, steel structured roofs, wooden roofs, etc., and covers the building exterior walls with vegetation. Research results show that the indoor air temperature of buildings with wall greenery is about 3 °C lower than that of indoor without wall greenery, demonstrating a significant cooling effect.

In summary, wall energy-saving insulation technology employs materials with thermal insulation and energy-saving properties, and excellent architectural functionality. Through advanced construction techniques, these materials are affixed to building walls, effectively minimizing internal temperature fluctuations, enhancing energy efficiency, and reducing the consumption of fossil fuels for air conditioning, heating, and electricity use, thereby indirectly decreasing carbon dioxide emissions.

d. Cool Roofing

Cool roofing is an innovative roofing design that adheres to principles of sustainability, environmental protection, and low carbon emissions. It achieves energy conservation and emission reduction by reflecting heat and sunlight. This technology employs reflective coatings and specialized roofing materials to reduce heat absorption and reflect thermal radiation, assisting houses and buildings in maintaining standard indoor temperatures, reducing reliance on air conditioning, and lowering the consumption of energy such electricity, thereby indirectly reducing carbon dioxide emissions from power plants.

e. Electrochromic Smart Glass

Electrochromic smart glass is an emerging technology and a sustainable building solution that can effectively block the heat of solar radiation. This smart glass can be integrated with building control systems, utilizing tiny electrical signals to achieve micro-charging of the glass, allowing users to adjust the amount of solar radiation reflection according to their needs. By employing this technology, residential and commercial buildings can significantly save the costs of heating, ventilation, and air conditioning. Currently, this technology is still being improved, and may be widely applied in the design and construction processes of sustainable buildings in the future as an intelligent energy-saving technology.

Chapter 7
The Application of Carbon Dioxide Removal Technologies in Achieving Carbon Neutrality

Achieving carbon neutrality requires not only the reduction of carbon emissions through various means but also the strengthening of the sequestration of greenhouse gases by enhancing carbon sinks and developing technologies such as Carbon Capture, Utilization, and Storage (CCUS). CCUS technology helps reduce the concentration of greenhouse gases in the atmosphere by capturing the carbon dioxide from industrial emissions or the atmosphere, then utilizing it or storing it underground. Meanwhile, by enhancing the carbon sink capacity of natural ecosystems such as forests, grasslands, and wetlands, as well as through technologies like Bio-Energy with Carbon Capture and Storage (BECCS), the absorption and storage of carbon dioxide can be further enhanced. In addition, Direct Air Capture (DAC) technology removes carbon dioxide directly from the atmosphere, while the increase of ocean and soil carbon sink capacities also provides support for carbon neutrality. The combined use of these technologies requires policy support, technological innovation, and capital investment to ensure the achievement of carbon neutrality goals and drive the economy towards a more sustainable and low-carbon future.

1 Carbon Sink and Its Application in Achieving Carbon Neutrality

Carbon sinks refer to the natural or artificial systems that can absorb and store the carbon (greenhouse gases) that would otherwise be released into the atmosphere. Among these systems, forestry carbon sinks have become the most outstanding one in terms of carbon sequestration capacity due to their stability and ease of human control. Forests absorb carbon dioxide through photosynthesis and convert it into organic matters that are then stored in the biomass of trees and forest soils for a long time. Meanwhile, the ocean, as the largest carbon sink on Earth, absorbs and

C. Chen et al., *Sustainable Carbon Neutrality*,
https://doi.org/10.1007/978-981-95-1691-9_7

stores a large amounts of carbon dioxide through its vast water bodies and biological activities, serving as a carrier with the largest carbon sink capacity.

In addition to forestry and ocean carbon sinks, ecosystems such as grasslands and wetlands also possess carbon sequestration potential that should not be ignored. Although this book does not elaborate on these carbon sinks, they contribute to reducing the concentration of greenhouse gases in the atmosphere through vegetation growth and other biogeochemical cycling processes, playing a crucial role in addressing climate change and achieving the goal of carbon neutrality. Protecting and restoring these ecosystems and enhancing their carbon sink functions are vital for mitigating climate change.

1.1 Forestry Carbon Sinks

Forests, being the main part of the terrestrial ecosystem, is a huge natural carbon pool. In recent years, China has continuously advanced the land greening initiatives, resulting in an increase of forest area up to 231 million ha (3.465 billion mu). Although China has a large forest area, the structure of tree species and tree ages is not optimally balanced, indicating great potential for the overall improvement of forest quality. Facing with the "carbon peaking and carbon neutrality" goals, it is necessary to continuously improve forest quality, thus increasing the carbon sinks of the forest carbon pool.

Forestry carbon sinks refer to the absorption and sequestration of carbon dioxide in the atmosphere through various forestry operation and management activities, such as forest protection, wetland management, desertification control, afforestation and forest management, by leveraging the natural carbon storage function of forest ecosystems. As the largest carbon pool in terrestrial ecosystem, forests play a unique and important role in reducing the concentration of greenhouse gases in the atmosphere and mitigating global warming. For every cubic meter increase of the forest stock volume, an average of 1.83 tonnes of carbon dioxide is absorbed and 1.62 tonnes of oxygen is released. Wood and bamboo have a very strong carbon sequestration and storage capability, which meets the needs for green and low-carbon development. Through these activities, forests can stabilize and increase the carbon stocks in vegetation and soil. After being monitored, measured and verified, the carbon-sink capacity can be utilized as carbon credits in the carbon trading market.

Forestry carbon sink projects offset the greenhouse gas emissions of enterprises through market-based means, generating additional economic values, including forest management carbon sink and afforestation carbon sink. Between them, forest management carbon sink focuses on existing forests. Through forest management measures, it promotes the growth of trees to increase carbon sinks. The afforestation carbon sink projects are jointly developed by the government, departments, enterprises and forest tenure entities. The government mainly plays a leading and guiding role. The forestry and grassland management departments are responsible for the organization of project development. Enterprises undertake the work such as carbon sink measurement, verification and market listing. Forest tenure entities are the ones who receive the benefits, as enterprises with greenhouse gases emission needs purchase carbon sinks. Forestry carbon sink projects not only help to improve the forest coverage rate and biodiversity but also enhance the adaptability of forest ecosystems to climate change.

Carbon sink trading is a market-based environmental and economic means aimed at reducing greenhouse gas emissions and increasing carbon sinks through market mechanisms. It allows project operators capable of reducing carbon emissions or increasing carbon sinks to obtain economic returns by selling carbon credits. These carbon credits represent the amount of carbon dioxide avoided or absorbed by the projects.

China's exploration in forestry carbon sink projects started in the context of international carbon trading. In 2004, the National Carbon Sink Management Office launched forestry carbon sink pilot projects in Guangxi, Inner Mongolia, Yunnan, Sichuan, Shanxi, Liaoning and other places. These projects increased the carbon storage capacity of forests by implementing activities such as forest protection and afforestation.

As the gradual establishment of domestic carbon trading market, carbon sink trading has been incorporated into China Certified Emission Reduction (CCER) trading system. The relevant methodologies for forestry carbon sink projects have also been put on record one after another, providing norms and basis for carbon sink trading.

Compared with other types of emission reduction projects such as wind power and photovoltaic, forestry carbon sink projects can not only reduce carbon emissions but also brings ecological and environmental benefits such as biodiversity protection, soil and water conservation. Therefore, they often have higher prices in the carbon market.

Through carbon sink trading, forestry carbon sink projects can secure necessary funding to promote more forest protection and afforestation activities, while bringing economic incentives to all parties involved in the projects. This market-oriented mechanism encourages participation from all sectors of society into carbon emission reduction and carbon sink increase efforts, facilitating the achievement of the carbon neutrality goal.

a. Types of Forestry Carbon Sink Projects

The domestic forestry carbon sink market mainly includes CCER (China Certified Voluntary Emission Reduction) forestry carbon sink projects, CGCF (China Green Carbon Sink Foundation) forestry carbon sink projects and regional-level forestry carbon sink projects. Compared with international forestry carbon sink projects, the review standards, technical standards and trading rules of these forestry carbon sink projects are more in line with my country's actual situation.

(i) CCER Forestry Carbon Sink Projects

CCER forestry carbon sink projects are an important component of the China Certified Emission Reduction (CCER) program. The main goal of these projects is to increase carbon sinks through forestry activities and help the country achieve the carbon neutrality goal. CCER forestry carbon sink projects include four major types: afforestation for carbon sinks, forest management carbon sink, bamboo afforestation carbon sink and bamboo management carbon sink.

Afforestation for carbon sinks and bamboo afforestation carbon sink projects fall into the category of afforestation carbon sink. Usually, they are initiated by the government and jointly implemented by departments, enterprises and forest tenure entities. These projects expand the forest coverage and enhance the absorption capacity of the ecosystems by planting more trees and bamboo. Afforestation carbon sink projects pay special attention to the benefits of forest tenure entities. The revenues generated from carbon sinks belong to the forest tenure entities, which motivates them to actively participate in afforestation activities and contribute to increasing the scale of carbon sinks. Forest management carbon sink and bamboo management carbon sink are classified under the category of forest management carbon sink. These projects promote the growth of existing forests and bamboo forests and improve their carbon storage capacity through scientific forest management measures such as thinning, selective cutting and forest regeneration. Compared with afforestation carbon sink projects, forest management carbon sink projects focus more on improving the quality and the carbon sink potential of existing forest resources.

Overall, CCER forestry carbon sink projects, through diversified forestry activities, not only enhance the ecological service functions of forests but also bring economic benefits to the project participants, thus promoting the sustainable development of forestry. The successful implementation of these projects is of great significance for facilitating the development of China's forestry carbon sink trading market and achieving China's carbon emission reduction and carbon neutrality goals.

(ii) CGCF Forestry Carbon Sink Project

The CGCF forestry carbon sink project, established by the China Green Carbon Foundation (CGCF) in July 2010, represents for a model of carbon sinks activities initiated by non-government sectors themselves. Operating in the form of a foundation, it mainly focuses on the field of afforestation carbon sink, providing a platform for enterprises, volunteers and all environmentally conscious individuals to participate in tree planting, forestry protection and forest-related public benefit activities.

The core objective of the CGCF project is to facilitate the protection and sustainable development of forest resources while increasing the carbon sink capacity of forests. In this manner, this project not only contributes to mitigating climate change but also supports biodiversity conservation and rural development. As of 2019, the CGCF foundation has established specialized green carbon funds in over 20 provinces, autonomous regions, and municipalities across the country, initiating more than 30 forestry carbon sink projects. The implementation of these projects has resulted in the creation of over 100,000 ha of afforestation for carbon sinks, significantly enhancing the carbon absorption and storage capacity of China's forests.

The success of the CGCF project demonstrates the potential and role of non-governmental organizations in advancing national carbon neutrality goal. Meanwhile, it also provides ways for other organizations and individuals to engage in national environmental protection and climate change mitigation efforts. Through such efforts, the CGCF foundation has made significant contributions to forest carbon sinks in China and globally.

(iii) Regional-Level Forestry Carbon Sink Projects

Regional-level forestry carbon sink projects are designed to benefit specific regions. They usually have local characteristics and are tailored based on specific circumstances to meet the needs of the local areas. These projects reflect China's innovation and diversity in the field of forestry carbon sinks, such as the BCER (Beijing Environment Exchange) mechanism in Beijing, the FFCER (Fujian Forestry Certified Emission Reductions) mechanism in Fujian Province, the Guangdong Tan Puhui pilot program, and the Individual Plant Carbon Sink Trading for Guizhou's Targeted Poverty Alleviation project. The main market targets of these projects are the local carbon trading markets. They meet the emission reduction needs of local enterprises and other participants by selling carbon credits. The revenues from these projects are mainly used for the protection and development of local forest resources, including afforestation, forest management, and biodiversity protection, with the aim of enhancing the carbon sink capacity of local forests, promoting the development of the local economy and the improvement of the ecological environment.

Although regional-level forestry carbon sink projects are independent of the national forestry carbon sink market, they provide a beneficial supplement to the country's overall carbon neutrality goal. Through these projects, local governments can respond more flexibly to the challenges of climate change and promote local green development and the construction of ecological civilization. Furthermore, these projects contribute to raising public awareness regarding the importance of climate change and forest protection, encouraging participation from all sectors of society in the efforts to reduce carbon emissions and increase carbon sinks.

In summary, regional-level forestry carbon sink projects are an important part of China's multi-level carbon market system. They not only contribute to the achievement of local sustainable development goals but also play a role in the national carbon neutrality strategy. Through these projects, China has provided valuable experience for the global response to climate change in exploring forestry carbon sink development modes tailored to the unique characteristics of different regions.

b. Characteristics of Forestry Carbon Sink Projects

(i) Reliance on the Carbon Sequestration Function of the Ecosystems Imposes Strict Requirements for Land Use

These projects absorb and store carbon dioxide in the atmosphere through the natural process of forests, namely photosynthesis, and convert carbon into organic matters that are accumulated in vegetation and soil. Since the development and implementation of forestry carbon sink projects must be based on the actual forest land, it is required that the project land must be suitable for forestry use and remain stable throughout the entire project cycle.

In China, the land use management system is highly stringent. Any change in land use or change in the management modes requires government review and approval. Forest management carbon sink projects rely on existing forest resources; therefore, the land on which the projects depend must be designated as forestry land and there should not be significant changes during the project cycle. Afforestation carbon sink projects focus on the establishment, restoration and improvement of forest vegetation. This requires that the project land should not only be suitable for forestry development but also have a long-term stable usage period to ensure that it can cover the project's development and operational cycle.

(ii) Widely Recognized and Supported by the International Community

Through afforestation and reforestation activities, these projects enhance the carbon sequestration capacity of the forests to mitigate climate change. In multiple international climate conventions, including the United Nations Framework Convention on Climate Change, the Kyoto Protocol, and the Paris Agreement, these activities are clearly listed as effective ways to reduce greenhouse gas emissions and increase carbon sinks.

The international community generally believes that forestry carbon sink projects not only help slow down the pace of global climate change but also play an important role in protecting and restoring biodiversity, improving regional climate, and promoting the sustainable management of water resources. In addition, these projects are also in line with the sustainable development goals of many countries, providing a viable path for achieving a win–win situation between environmental protection and economic development. International recognition also means that forestry carbon sink projects can be traded in the international carbon market, enabling project developers to obtain economic returns by selling carbon credits, which further motivates forest protection and restoration worldwide. In this way, forestry carbon sink projects not only contribute to the achievement of global climate goals but also bring economic and social benefits to the project participants.

(iii) Receiving Dual Supports from Policies and the Market in China

This feature reflects China's determination to promote green and low-carbon development and to achieve the goal of carbon neutrality. The Chinese government supports the development of forestry carbon sink projects through a series of policies and market incentive mechanisms, such as the China Certified Emission Reduction (CCER) program and the China Green Carbon Foundation (CGCF) project. These policies and mechanisms not only provide standardized management and certification processes for forestry carbon sink projects but also offer clear economic incentives for project developers.

Carbon credits generated by forestry carbon sink projects are allowed to be traded in the carbon market, providing direct economic returns to enterprises or individuals engaged in these projects. In this way, forestry carbon sink projects can attract more social capital investment to promote the protection and rational utilization of forest resources. Meanwhile, the operation of the carbon market also provides a variety of options for enterprises that need to reduce emissions, allowing them to offset their own carbon emissions by purchasing carbon credits, thus achieving the emission reduction goals at a lower cost.

In addition, support from policies and the market is also helpful for raising public awareness of climate change issues and encouraging active participation from all sectors of society in the carbon emission reduction and carbon neutrality efforts. The implementation of forestry carbon sink projects can promote the restoration and protection of forest ecosystems, improve the ecosystem service functions, and contribute to the achievement of the country's ecological construction and sustainable development goals at the same time.

(iv) Multiple Benefits Along with Great Uncertainty

These projects can not only slow down climate change by absorbing the carbon dioxide in the atmosphere, but also generate ecological, economic and social benefits such as biodiversity protection, soil and water conservation and the provision of recreational space. However, the revenues from forestry carbon sink projects exhibit a relatively high uncertainty, mainly because of the long project cycle. Forestry projects usually take many years to reach maturity, and during this period, they may be affected by various factors such as natural conditions, market fluctuations and policy adjustments. From an investment perspective, longer project cycle brings higher uncertainty and risks, which leads to an increase in the uncertainty of revenues.

For example, the cycle of international forestry carbon sink projects such as the Clean Development Mechanism (CDM), the Verified Carbon Standard (VCS) and the Gold Standard (GS) is usually more than five years, and even more than ten years for some projects. In China, the cycle of forestry carbon sink projects is generally around three years. Even for projects with relatively shorter cycle, such as the Fujian Forestry Certified Emission Reductions (FFCER) and the Guangdong Province Puhui Certified Emission Reductions (PHCER) program, they also require more than one year. The high uncertainty of returns brought about by such long cycles may affect the motivation of investors and project developers. To mitigate risks, project developers need to conduct meticulous project planning and risk assessment, and may need to seek for policy support, insurance mechanisms or other risk mitigation measures. Meanwhile, policymakers can encourage and promote the development of forestry carbon sink projects by providing incentives and a stable policy environment, such as tax preferences and financial subsidies. These measures can reduce the uncertainty related to forestry carbon sink projects to a certain extent, thus enhancing their contributions to climate change mitigation and sustainable development.

c. Key Points for the Development of Forestry Carbon Sink Projects

(i) Requirements of Methodologies

Methodology serves as a methodological guide for establishing the project baselines, demonstrating additionality, calculating emission reductions, and formulating monitoring plans, etc. It plays a crucial role in the development of forestry projects. During the first CCER period (from 2013 to 2017), the National Development and Reform Commission issued seven methodologies related to forestry carbon sinks (which are currently invalid); in the new CCER period (starting from 2023), the Ministry of Ecology and Environment has released two forestry carbon sink methodologies regarding afforestation for carbon sinks and mangrove vegetation creation, which have already come into effect. Relevant methodologies generally follow the principles of conservativeness, transparency, comparability, reduction of uncertainty, and cost-effectiveness. They set strict regulations on the conditions for applicability, land eligibility, project boundaries, baselines and additionality, as well as carbon sink measurement and monitoring. It is worth noting that due to the requirement for additionality, the existing forest carbon stock is not equivalent to carbon-sink capacity. Afforestation for carbon sink project is a process of creating or restoring forests through artificial measures on planned plantations that do not meet the definition of a forest. No matter what type of carbon sink project is being developed, the requirements of the methodology should be strictly implemented.

(ii) Determination of Ownership and Property Rights

Forestry carbon sinks, as a tradable commodity, will inevitably involve issues of ownership and property rights. Clear land ownership and clear forest ownership are important prerequisites for the development of forestry carbon sink projects. Currently, the common practice is that project owners must provide land ownership certificates issued by the people's governments at or above the county level in the early stage of project development. If the project owner does not directly possess the land ownership or the forest usage right, it can sign relevant agreements with the actual rights holders under the circumstances permitted by laws, to entrust the development of the project. However, this does not mean that some intermediary agencies can bypass the owners of forest land and forests to directly sign agreements with local governments and obtain a share of the project benefits through "cooperative" development and other means. Such actions contravene the property right attributes of forestry carbon sinks as a commodity.

(iii) Project Development and Management

The development of forestry carbon sink projects in China mainly refers to the processes established by international carbon emission reduction project mechanisms, such as the United Nations Clean Development Mechanism (CDM). According to "Measures for the Administration of Voluntary Greenhouse Gas Emission Reduction Trading (Trial)", China adopts a registration management system for the voluntary greenhouse gas emission reduction trading, including project registration and certified emission reductions (CERs) registration. The project development and trading process usually includes: project development assessment, project design and development, project validation and registration, project implementation and monitoring, verification and registration of emission reductions, and trading of emission reductions. Compared with other new energy and renewable energy projects, the development and management of forestry carbon sink projects are characterized by lengthy emission reduction output cycle, complex monitoring, and high development costs. This also leads to their relatively scarcity in CCER projects and emission reductions registrations.

(iv) Case Study of Forestry Carbon Sink Projects

a. China Certified Emission Reductions (CCER)

In 2015, the Guangdong Chimelong Afforestation for Carbon Sink Project, the first CCER forestry carbon sink project in China, received the record registration and issuance of emission reductions by the National Development and Reform Commission. In 2011, this project implemented afforestation for carbon sinks in three counties in Guangdong Province, covering an area of 8.7 km^2, with an expected emission reductions of 347,000 tonnes over a 20-year period. Currently, the total emission reductions from this project, including 5208 tonnes of CCER carbon sinks issued in the first phase, have been contracted for purchase by the Guangdong Yudean Group Co., Ltd., an enterprise subject to emission control, at the unit price of 20 yuan per tonne for fulfilling its obligations. The successful development and trading of this project have provided valuable experience and a good example for China's CCER forestry carbon sink projects.

b. CGCF Rural Households' Forest Management Carbon Sink Trading Project

To promote forest management after the reform of collective forest tenure system and increase the income of forest farmers, the China Green Carbon Foundation and Zhejiang A&F University developed the "Trading System for Rural Households' Forest Management Carbon Sink Projects" in 2014. This system, which follows international rules, is tailored based on China's specific conditions and the characteristics of decentralized forest management by rural households after the reform of forest tenure system, as well as current domestic and international policies and practices regarding voluntary carbon trading. It took the forest management of rural households in Lin'an District, Hangzhou, Zhejiang Province as a pilot project and established a forest management carbon sink trading system that covers project design, auditing, registration, issuance, trading, and supervision. It specifies the administrative roles of government departments, with scientific research departments providing technical services, and a third party conducting project validation and verification, and registration to ensure the real existence of carbon sequestration and emission reductions. Finally, the projects are entrusted to the East China Forestry Exchange for trading.

The first phase involved 42 rural households whose forest management carbon sink projects generated an emission reduction of 4285 tonnes, which was purchased by the Zhejiang branch of China Construction Bank to offset the annual carbon emissions of its office buildings, thereby achieving carbon neutrality for its office operations. This marks the first monetary return that farmers have received from forest carbon sink trading since the implementation of China's forest tenure reform. Although the transaction volume is not high, it establish a valuable precedent for promoting the trading of forest ecological services.

c. Regional-Level Forestry Carbon Sink Projects

The Guangdong Provincial Development and Reform Commission has issued the "Implementation Plan for the Guangdong Tanpuhui Pilot Program" and the "Interim Measures of the Development and Reform Commission of Guangdong Province on the Management of Puhui Certified Emission Reductions (PHCER)". On November 28, 2019, a bidding transaction for forestry PHCER projects in 24 provincially designated poverty-stricken villages across three counties (cities, districts) including Shixing County, Shaoguan was held at the Guangzhou Carbon Emission Trading Exchange, with participation from six organizational and individual members. The total emission reductions from the forestry PHCER projects of these 24 provincially designated poverty-stricken villages in three counties (cities, districts) including Shixing County, Shaoguan amounted to 196,643 tonnes, and the transaction was made at RMB 32.02 yuan per tonne.

1.2 Ocean Carbon Sinks

The ocean, as the largest active carbon pool on Earth, plays a crucial role in addressing climate change. It sequesters the majority of the planet's carbon dioxide and absorbs and removes a large amount of CO_2 emitted from human activities every year, demonstrating its central position in the carbon cycle and global climate regulation. The carbon storage cycle of ocean carbon sinks is considerably longer than that of terrestrial ecosystems. It can sequester carbon for hundreds or even thousands of years, which makes the ocean a long-term and stable carbon sink in the global carbon cycle.

China, with its extensive coastline and rich marine ecological resources, possesses immense potential for the development of ocean carbon sinks. However, compared to forestry carbon sinks, the development and management of ocean carbon sinks in China are still in a primary stage. Currently, the rules, regulations and industry standards related to ocean carbon sinks are not fully developed, and the ocean carbon sink market has yet to be fully established. To fully tap the potential of ocean carbon sinks and promote their development, the following measures can be implemented.

(1) Strengthen the research on ocean carbon sinks to improve the understanding of the mechanism and capacity of ocean carbon storage.
(2) Formulate policies and standards related to ocean carbon sinks to provide guidance for the development and management of ocean carbon sink projects.
(3) Promote the development of ocean carbon sink projects, such as the establishment of marine protected areas, the restoration of mangroves, and marine ecological restoration, to enhance the carbon sequestration functions of marine ecosystems.
(4) Explore the inclusion of ocean carbon sinks in the carbon trading market to create a synergy effect with forestry carbon sinks, etc., and build an integrated carbon sink system.

(5) Improve the understanding of the public and stakeholders regarding the importance of ocean carbon sinks, and encourage participation from various sectors in the protection and development of ocean carbon sinks.

a. Characteristics of Ocean Carbon Sinks

The characteristics of ocean carbon sinks are demonstrated in the fact that, as the largest active carbon pool on Earth, they possess strong carbon absorption and storage capabilities. However, human activities such as coastal eutrophication, land reclamation, coastal engineering, and coastal urbanization have adversely affected ocean carbon sinks, resulting in a decline in their carbon sequestration capacity. Therefore, the key to restoring and enhancing ocean carbon sinks lies in the restoration and protection of the marine ecosystem.

Marine aquaculture is one of the effective ways to increase ocean carbon sinks. By developing the cultivation of aquatic products such as shellfish and algae, the absorption of phytoplankton by these organisms can be utilized to capture carbon elements in seawater. When these aquatic products are harvested and removed from the ocean carbon cycle, carbon sequestration is achieved. In addition, the cultivation of seaweed and seagrass contributes not only to environmental optimization but also to the sustainable supply of marine energy. The protection of coastal wetlands is another important mode of increasing carbon sinks. By formulating and implementing specific policies to protect the ecological environment of coastal wetlands such as seagrass meadows, salt marshes, and mangroves, the carbon sink functions of these ecosystems can be fully exerted. These wetland ecosystems play a key role in carbon sequestration and are an important part of ocean carbon sinks. Pollution emission control is another key measure to restore the carbon sink capacity of the ocean. Reducing pollutant emissions can help the marine environment restore its natural state and enhance its carbon sink capacity. This includes reducing the inflow of terrestrial pollutants into the ocean and controlling the impact of offshore activities on the marine environment.

Increasing ocean carbon sinks requires a comprehensive approach that integrates ecological restoration, marine aquaculture, coastal wetland protection, and pollution emission control. Through these measures, we can enhance the ocean's carbon sequestration capacity, fight climate change, and promote the healthy and sustainable development of the marine ecosystems.

b. The Development Pathay of Ocean Carbon Sinks

The development path of ocean carbon sinks is a comprehensive and complex systematic endeavor that requires comprehensive consideration and coordination in multiple aspects from ecological protection, ecological restoration, pollution control to market-based mechanism. The core of this path lies in achieving the healthy and sustainable development of the marine ecosystems through scientific planning and effective management.

(i) Ecological Restoration

The key to ecological restoration lies in the restoring and protecting of the key ecosystems that are critical for carbon storage. Ecosystems such as mangroves, seagrass beds, and coastal wetlands, despite their relatively small area, play a vital role in carbon sequestration. These ecosystems absorb and store significant amounts of carbon dioxide through natural processes, making them irreplaceable in mitigating global climate change. To protect and enhance the carbon sink functions of these ecosystems, a coordinated mechanism integrating land and sea must be established. This mechanism can reduce the negative impacts of human activities on ocean carbon sinks. For example, by reducing coastal eutrophication, controlling land reclamation and coastal engineering, and other human disturbances, the damage to marine ecosystems can be minimized. By adopting this approach, the health of the ecosystems can be maintained and enhanced, and their resilience to climate change and carbon sink capacity can be strengthened. Furthermore, ecological restoration also includes the restoration of degraded ecosystems, such as replanting vegetation, restoring natural hydrological conditions, and controlling invasive species. These measures help to restore the structure and function of the ecosystems, thereby improving its carbon sequestration capacity. Ecological restoration not only contributes to the enhancement of biodiversity and ecosystem services, but also serves as an important means to achieve regional sustainable development and address climate change.

(ii) Ecological Compensation Mechanism

The ecological compensation mechanism plays a crucial role in the development of ocean carbo sinks. It motivates the engagement of various groups in the protection and construction of ocean carbon sinks by offering economic incentives. The core of this mechanism is to ensure that the protection and restoration of marine ecosystems can receive financial support and can make advancements in scientific research and technologies. Economic incentive measures can include direct financial subsidies, preferential taxation, reward mechanisms, etc., aiming to encourage individuals, communities, enterprises, and government agencies to participate in marine protection projects. These incentives help to raise awareness of the importance of ocean carbon sinks among all parties and encourage their active participation in practices. At the same time, the ecological compensation mechanism also emphasizes support for scientific research projects. By funding relevant researches, it promotes in-depth understanding of the ocean carbon sink function and the methods for its enhancement. This includes research on the carbon sequestration capacity of marine ecosystems, the carbon cycle process, and the environmental and social benefits of carbon sink projects.

Additionally, the establishment of an ecological information monitoring system is a vital component of the ecological compensation mechanism. Developing and improving this system enables effective monitoring, assessment, and management of ocean carbon sink projects, thus ensuring the implementation effectiveness and long-term sustainability of the projects. This involves the collection, analysis, and sharing of data, as well as regular reviews of project progress and impacts. Through the efforts made by stakeholders, the ecological compensation mechanism aims to create a comprehensive framework that promotes the protection and enhancement of ocean carbon sinks, thus contributing to addressing climate change and advancing sustainable development.

(iii) Ocean Carbon Sink Trading

Ocean carbon sink trading, as a market-based mechanism, plays a decisive part in promoting the development of ocean carbon sinks. By formulating appropriate trading rules within the framework of the existing carbon trading market, it effectively channels funds and resources towards the construction and protection of ocean carbon sinks. The implementation of this mechanism is a staged process, aiming to gradually integrate ocean carbon sinks into the core system of carbon emission governance.

In the initial stage, the focus will be on ecological compensation and scientific research to ensure an accurate understanding and assessment of the potential and value of ocean carbon sinks. This requires in-depth research on the carbon storage capacity of marine ecosystems and the establishment of corresponding scientific foundations and assessment methodologies. As the market grows, the focus will shift to the construction of the market system, including the improvement of the monitoring, assessment and trading processes for carbon sinks. This stage requires the establishment of transparent trading rules and regulatory mechanisms to ensure the fairness and effectiveness of ocean carbon sink trading. When entering the stage of international development, ocean carbon sink trading will expand to the global market. The existing systematic achievements will be utilized to actively participate in international climate governance and strive for greater international influences. This not only helps to enhance the status of ocean carbon sink projects in the global carbon market but also promotes the international community's awareness of the importance of ocean carbon sinks and the participation in this effort.

In addition, ocean carbon sink trading also involves exploring a market-based governance path based on carbon emission trading, encouraging wide participation from all sectors of society. By incorporating standardized and quantified ocean carbon sinks into the trading system, the whole society's awareness of the importance of ocean carbon sinks can be enhanced, thus stimulating more social participation.

The development of ocean carbon sink trading is a gradual process that requires the joint efforts in policy support, scientific research, market mechanisms, and international cooperation, to unleash the potential of ocean carbon sinks in addressing climate change and promoting sustainable development.

(iv) Assessment of Ocean Carbon Sinks

The assessment of ocean carbon sinks is crucial for ensuring that the ocean carbon sink projects are scientific and effective. It requires the establishment of a comprehensive and systematic monitoring and evaluation system. The establishment of this system requires increased investment in basic research on ocean carbon sinks, a deeper understanding of the carbon sequestration capacity of marine ecosystems, and additional ways to enhance their carbon sink potential. In addition, it is important to establish a national-level information sharing platform for ocean carbon sinks. This platform can integrate and share data from different sources, improving the transparency and accessibility of ocean carbon sink assessments. By facilitating the connectivity of multi-source data, the platform will support more accurate measuring, reporting, and verification (MRV) of ocean carbon sinks. At the same time, encouraging the engagement of scientific research institutions, government agencies, enterprises, and the public at home and abroad into the analysis and evaluation of ocean carbon sinks, thus harnessing collective wisdom and promoting knowledge sharing and technical exchanges. Such extensive participation not only helps to improve the quality and credibility of the assessment work but also facilitates the formation of a global collaboration network.

Improving the accuracy and reliability of assessments is the key to ensuring the success of ocean carbon sink projects. Accurate assessment results can provide a solid scientific basis for policy-making, project investment, and market transactions. This is crucial for the protection, development, and trading of ocean carbon sinks, helping to maximize the economic and ecological benefits of ocean carbon sink projects. Ultimately, through these comprehensive measures, ocean carbon sink assessments will support the promotion of the sustainable development and international cooperation of ocean carbon sinks, contributing to global climate actions. Through scientific assessment and effective management, ocean carbon sink projects can play a more important role in global carbon reduction and ecosystem protection.

2 Carbon Capture, Utilization and Storage Technologies

Carbon Capture, Utilization, and Storage (CCUS) technology refers to the process of separating carbon dioxide from emission sources, followed by its capture, direct utilization, or storage, to achieve carbon dioxide reduction. It mainly includes carbon capture, transportation, storage, and utilization technologies. CCUS is a comprehensive technological solution aimed at reducing the concentration of carbon dioxide in the atmosphere and mitigating climate change.

This technology involves separating CO_2 from industrial processes or energy use, which is then utilized or securely sequestrated through different methods to achieve permanent CO_2 emission reduction. The CCUS technology process includes the following links: capture, transportation, utilization, and sequestration. In the capture link, the separation of CO_2 is mainly achieved through pre-combustion, post-combustion, and oxy-fuel combustion processes. In the transportation link, CO_2 is transported to the destination through pipelines, ships, etc. In the utilization link, engineering and technological means are used to convert CO_2 into useful products or resources, including mineral carbonation and physical, chemical, and biological utilization. In the storage link, through geological or ocean sequestration technologies, CO_2 is isolated underground or in the deep sea for a long time to reduce emissions.

Despite the great potential of CCUS technology, its development and application still face challenges in terms of technologies, costs, environmental risks, and policies.

2.1 Carbon Capture

Carbon capture refers to the process of separating CO_2 from industrial production, energy utilization, or the atmosphere. It is mainly categorized into pre-combustion capture, post-combustion capture, oxy-fuel combustion, and chemical-looping capture. This process can be achieved in various ways, including pre-combustion capture, post-combustion capture, oxy-fuel combustion, and chemical-looping capture. Usually, it is easier to capture the CO_2 from industrial exhaust gas flow (point source carbon capture) due to its higher concentration of CO_2. Currently, most large-scale industrial carbon capture projects adopt this method. Nevertheless, direct capture of CO_2 from the atmosphere (direct air capture), while facing higher costs and technical challenges, has a greater potential to effectively reduce the CO_2 concentration in the atmosphere. With the development of technology and the reduction of costs, direct air capture technology is expected to play an increasingly important role in future carbon-reduction strategies.

a. Point Source Technology

In point source CCUS technology, CO_2 generated by industries such as cement and steel production, hydrogen production from fossil fuels, waste incineration, and power generation can be captured before it is emitted into the atmosphere. The captured CO_2 is then compressed to over 100 atmospheres and injected into porous rock layers more than 1000 m underground, where it can be stored for tens of thousands to millions of years beneath impermeable rocks. CO_2 can also be incorporated into products such as building materials for long-term storage. Through different engineering methods, CO_2 can be effectively captured from point sources, with a capture rate above 90%. The cost is 10–100 dollars per tonne of CO_2. Although the cost is higher than that of greenfield projects, the carbon capture equipment can be retrofitted from existing infrastructure, enabling the realization of a net-zero strategy.

(i) Cement

Traditional cement manufacturing involves heating carbonate feedstocks (usually limestone calcium carbonates) at high temperatures in a rotary kiln. High-temperature "calcination" generates calcium oxide and carbon dioxide. In addition, the combustion of fuels (usually coal or natural gas), which generates heat necessary for the above-mentioned conversion reactions, also produces carbon dioxide. Carbonates are a key component of Portland cement, which is widely used in the global construction industry. Even when bio-sourced or other low-carbon fuels are used for heating, the cement calcination reaction still accounts for approximately 50% of the emissions. These emissions are crucial for the chemical reactions that produce calcium oxide. The cement industry accounts for 8% of global CO_2 emissions, with calcination alone accounting for about 4%. Although there are alternative products to cement, their adoption has been slow. Therefore, addressing the emissions problem in the cement industry is of great importance for achieving a net-zero world. The flue gas from cement kilns is a good target for implementing CCS. The CO_2 concentration in the flue gas is generally 14–33%, which is higher than that of traditional coal-fired emissions. In addition, due to its high CO_2 purity, the energy intensity for the capture is lower. Otherwise, greater treatment efforts are required to remove pollutants such as cement dust.

(ii) Steel

The Emirates Steel plant in Abu Dhabi has been conducting carbon capture and storage using the solvent capture method since 2016. CO_2 comes from the coal or natural gas used as reducing agents in the Direct Reduced Iron (DRI) units, which convert iron ore into iron for steelmaking. The steel plant captures approximately 800,000 tonnes of CO_2 annually and transports it via pipelines for enhanced oil recovery (EOR). Other projects are also exploring changes to the basic steelmaking process to promote CO_2 emission reduction. Tata Steel's Hisarna process is a new technology which can not only improve energy efficiency, reduce the emission intensity of steelmaking, but also increase the CO_2 concentration, making it more conducive to capture.

b. Direct Air Carbon Capture and Storage (DACCS) that Integrates Carbon Sequestration

The method of directly capturing CO_2 from the atmosphere with DACCS facilities has several key advantages.

- The capture plant can be located in areas with abundant wind energy, thereby reducing the operating costs of wind turbines.
- The capture plant can be located in places with access to renewable electricity.

Capturing CO_2 from the atmosphere is more difficult than capturing it from other sources due to the low CO_2 concentration, which is approximately 0.04% (400 ppm) in the atmosphere. The CO_2 concentration in the atmosphere is only one-hundredth of that in the flue gas of a gas-fired power station. Compared with emission sources with higher CO_2 concentration, much more energy is consumed to concentrate CO_2 from the atmosphere with such a low CO_2 concentration.

Carbon Engineering, a Canadian company, has developed a DAC technology that uses a liquid potassium hydroxide solvent to absorb CO_2 from the atmosphere. Through a chemical reaction, CO_2 is captured and converted into carbonate, which is then extracted to form granules, and subsequently calcined (heated to separate CO_2) to generate pure CO_2. The energy required for this entire process comes from externally provided renewable electricity or the combustion of natural gas. If natural gas is used, the CO_2 released during combustion is captured and stored in the process, thus achieving negative emissions. This type of plant can operate flexibly using either electricity or natural gas.

Oxy Low-Carbon Ventures, a subsidiary of Occidental Petroleum, plans to build a commercial-scale DAC plant and use the process developed by Carbon Engineering. Climeworks in Sweden and Global Thermostat in the United States have adopted different DAC methods. Their technologies rely on patented solid-state absorption materials to adsorb CO_2 from the air. Once the adsorbent is saturated with CO_2, it is heated to desorb the CO_2. This is a form of the temperature swing adsorption (TSA) process. The temperature swing adsorption process has a long history of industrial application, but this is its first application in DAC. The TSA processes of these two companies also capture a large amount of water from the atmosphere, thus producing clean water as a useful by-product.

2.2 Carbon Utilization

Carbon utilization is a process in CCUS technology that converts captured CO_2 into useful products or resources. It covers multiple aspects such as mineral carbonation, physical utilization, chemical utilization, electrochemical conversion, biological utilization, geological sequestration utilization, deep-sea sequestration, and applications in the construction industry. The utilization of CO_2 provides a way to transform greenhouse gases into useful products, which range from fuels, chemicals, building materials to polymers, etc. Although the captured CO_2 is currently mainly used for underground storage or enhanced oil recovery (EOR), its potential as an industrial feedstock is huge. CO_2 can be chemically converted into various types of products, such as fuels and energy products like methanol, methane, and syngas, etc. These products may release carbon rapidly after being used, but they are part of the carbon neutrality strategy. As a building material, CO_2 can form carbonates through chemical reactions, thus solidifying CO_2. For example, calcium carbonates are formed during the cement production process, which not only helps reduce CO_2 emissions but also improves the durability of concrete. As a chemical product, CO_2 can be converted into various chemical products, such as polycarbonates, organic carbonates, and urea, etc. Although these conversion processes usually require catalysts and specific reaction conditions, they offer extensive application prospects for the utilization of CO_2 as a resource. As a biomaterial, CO_2 can be converted into biomass or biofuels using microbial conversion technology. Usually, through a specific microbial fermentation process, CO_2 is transformed into valuable biobased products.

However, CO_2 is a very stable molecule, so the process of converting CO_2 into useful products requires a high energy input, which limits its wide application at a certain level. To address this challenge, innovative companies around the world are developing new technologies to improve the energy efficiency of CO_2 conversion. With the increasing availability of cheap renewable energy, the utilization of CO_2 is gradually emerging as a commercially viable industry. The progress of these technologies and the reduction in energy costs are expected to drive the development of CO_2 utilization technologies, making them one of the effective ways to reduce greenhouse gas emissions and achieve the carbon neutrality goal.

2.3 Carbon Storage

The storage of CO_2, commonly known as carbon sequestration, is a crucial part of the Carbon Capture, Utilization, and Storage (CCUS) technology process. This technology involves the long-term and safe storage of captured CO_2 to prevent its release into the atmosphere, which contributes to the mitigation of climate change. The sequestration of CO_2 can be achieved through various means, including geological sequestration, land surface sequestration, and ocean sequestration.

Geological sequestration is a technology that safely stores CO_2 deep underground for a long period of time. By mimicking the process of fossil fuel storage by nature, it injects CO_2 into specific geological formations. These geological environments, such as depleted oil and gas fields, inaccessible coal seams, deep saline aquifers, etc., must have suitable structures such as caprocks, reservoirs, and traps, to ensure effective CO_2 sequestration and prevent its leakage back to the land surface or migration to other areas. Under such conditions, CO_2 can be safely sequestered underground for thousands or even tens of thousands of years, thus reducing the amount of greenhouse gases in the atmosphere and combating climate change. Worldwide, multiple CO_2 geological storage projects are underway. These projects not only demonstrate the technical feasibility but also provide practical solutions for the achievement of carbon reduction and carbon neutrality goals. With the development of technologies and the accumulation of experience, geological sequestration is expected to become one of the key technologies in the global response to climate change.

Land surface sequestration is a long-term and stable method for CO_2 sequestration by converting it into solid carbonates. This process involves a chemical reaction between CO_2 and metal oxides. The resulting carbonate minerals, due to their stability, can safely store CO_2 on a geological timescale of over a thousand years. This sequestration method offers the advantage of low regulatory costs as it reduces the need for long-term monitoring. However, the land surface sequestration technology is not yet fully mature and faces some challenges. The operational costs are relatively high because this process requires a large amount of energy and feedstocks. In addition, environmental impact considerations need to be carefully taken into account, including the handling of by-products related to the chemical reactions and the possible impact on the local ecosystems.

Ocean sequestration is a method for long-term CO_2 isolation, which takes advantage of the vast volume of the ocean and the solubility of CO_2 in water. By injecting CO_2 into the deep sea, these gases can exist in a liquid or supercritical state under the high-pressure and low-temperature conditions of the deep sea, thus being isolated from the atmosphere for possibly hundreds or even thousands of years. The advantage of this sequestration method is that the ocean has a huge potential sequestration capacity which is estimated to be much larger than the current CO_2 volume in the atmosphere. However, the implementation of ocean sequestration requires careful consideration of its impact on the marine ecosystem. An increase in CO_2 concentration in seawater may lead to ocean acidification, affecting the calcification process, the reproduction, growth, and migration abilities of marine organisms, and thus having a negative impact on the entire marine ecosystem. To ensure the safety and effectiveness of ocean sequestration, rigorous scientific research and environmental impact assessments are needed. This includes monitoring the dissolution and dispersion process of CO_2 in seawater, evaluating its long-term impact on the marine environment, and formulating strict technical specifications and operational standards.

In China, preliminary progress has been made in the research and engineering demonstration of CO_2 geological sequestration technology. For example, the Ordos CO_2 geological sequestration demonstration project has showcased the feasibility and geological safety of this technology. In addition, compressed CO_2 energy storage (CCES), as a new energy storage technology, features high cycling efficiency, low construction complexity, long operational lifespan, and low system costs. Moreover, it can be closely integrated with carbon capture technology, providing new ideas for achieving the carbon peaking and carbon neutrality goals.

MIX
Papier aus verantwortungsvollen Quellen
Paper from responsible sources
FSC® C105338

If you have any concerns about our products,
you can contact us on
ProductSafety@springernature.com

In case Publisher is established outside the EU,
the EU authorized representative is:
Springer Nature Customer Service Center GmbH
Europaplatz 3, 69115 Heidelberg, Germany

Printed by Libri Plureos GmbH
in Hamburg, Germany